BEIHEFTE ZUM TÜBINGER ATLAS
DES VORDEREN ORIENTS

herausgegeben im Auftrag des Sonderforschungsbereichs 19
von Helmut Blume und Wolfgang Frey

Reihe A
(Naturwissenschaften)
Nr. 24

Harald Kürschner (Ed.)

Contributions to the Vegetation of Southwest Asia

WIESBADEN 1986
DR. LUDWIG REICHERT VERLAG

Contributions to the Vegetation of Southwest Asia

by

Harald Kürschner (Ed.)

WIESBADEN 1986
DR. LUDWIG REICHERT VERLAG

CIP-Kurztitelaufnahme der Deutschen Bibliothek

Contributions to the vegetation of Southwest Asia / [diese Arbeit ist im Sonderforschungsbereich 19, Tübingen entstanden]. Harald Kürschner (ed.). – Wiesbaden : Reichert, 1986.
(Beihefte zum Tübinger Atlas des Vorderen Orients : Reihe A, Naturwissenschaften ; Nr. 24)
ISBN 3-88226-297-4

NE: Kürschner, Harald [Hrsg.]; Sonderforschungsbereich Orientalistik <Tübingen>; Tübinger Atlas des Vorderen Orients / Beihefte / A

Gesamtherstellung: Hubert & Co., Göttingen. Printed in Germany

Preface

With its rich diversity of land forms, its pronounced topographical, orographical, pedological, geomorphological and climatic contrasts and its resulting variety of biogeocenoses, south-west Asia is one of the most fascinating regions of the Holarctic Kingdom. Initial geobotanic research on this region understandably focused on the northern parts (the Levant, Turkey and portions of Iran and Afghanistan), which were of greater cultural interest from the Western point of view, besides being more densely settled and accessible. Today fairly good information is available about the flora and vegetation of these areas. In contrast, large parts of south-west Asia were virtually ignored for many years on account of political obstacles, strenuous travel conditions or supposed unattractiveness. One such area that was largely neglected by geobotanists is the Arabian Peninsula. In the last five years, however, there has been a marked acceleration of exploration and floristic survey work here, and the first scientifically founded results are now available.

In the present volume a deliberate effort has therefore been made to include a number of studies concerned with vegetation conditions in the Arabian Peninsula. Detailed investigations as well as summary articles are presented on Jordan, Saudi Arabia, Kuwait, North Yemen and Oman, countries which until now have been dealt with only in accounts of journeys or in publications of a very special or very general nature. Consequently this book can be regarded as an introduction to vegetation conditions in the Arabian Peninsula and as a starting point for those who wish to gain a better acquaintance with the region. The articles are also intended to explain and to supplement the general maps of vegetation in south-west Asia (Maps A VI, 1,2) that have been published in the Tübingen Atlas of the Middle East (TAVO). Because of their large scale (1 : 8 million) these maps are unable to provide detailed information and to depict local circumstances.

The volume is rounded out by a recent survey of vegetation conditions in Iran. Up to now a study of this type has been lacking.

The Editor hopes that this book will give a picture of the current state of geobotanic research on the region and that it will enable readers to deepen and improve their knowledge. An additional goal is to stimulate further research on south-west Asia and to make this region, which is of such interest in geobotany, known to a larger circle.

Finally the Editor wishes to extend his thanks to the writers of the various papers for their willingness to contribute manuscripts. He also wishes to thank the Deutsche Forschungsgemeinschaft, Sonderforschungsbereich 19, "Tübinger Atlas des Vorderen Orients", and especially its spokesman Prof. Dr. W. Röllig, for making this publication possible.

Harald Kürschner — Berlin, März 1985

Contents

Wolfgang Frey and Wilfried Probst

A synopsis of the vegetation of Iran

Contents

1. Introduction

Of all the countries of south-west Asia Iran has the most varied and most fascinating vegetation conditions. Today one still finds humid forests in parts of the south Caspian coastal plain; humid forests dominate the northern slopes of the Elburz Mountains, and drought-adapted woodlands cover the mountains of Kurdistan, the Zagros ranges and the ranges of northern Ḫorāsān. In the subalpine zones of these ranges *Thorn-cushion formations* are dominant. The arid parts of the country are occupied by *Dwarf-scrub formations* and *Halophytic communities*, and the coastal regions on the Persian Gulf are characterized by *Semidesert shrublands* belonging already to the Paleotropic kingdom.

The variety of Iran's flora and vegetation is attributable to the country's large climatic contrasts and is also the result of the area's floral history and evolutionary potential. Numerous plant taxa have their centres of diversity in the Irano-Afghan region. The mountain regions are particularly rich in neoendemites. Stocks of the genera *Astragalus*, *Acantholimon*, *Acanthophyllum*, *Nepeta*, *Onosma* and *Cousinia* are examples. The variety of the flora and vegetation is furthermore enriched by refugia of floras that once had a wide distribution.

This study gives a survey of the vegetation of Iran with regard to its phytogeographical status, its basic features and its present-day structure.

2. Short history of investigation

Since the 16th century numerous researchers have collected and described plants in south-west Asia. E. Kaempfer was one of the first to investigate the flora of Iran (1664–68). Other early names are S. G. Gmelin and P. S. Pallas (1770–72), A. Michaux (1782), G. A. Olivier and J. G. Bruguière (1796–97), P. M. R. Aucher-Eloy (1835–38) and T. Kotschy (1841–42). The first milestone in floristic research on south-west Asia, and thus on Iran, was Boissier's "Flora Orientalis" (1867–1888). The second outstanding contribution was made by Bornmüller, who through his travels and numerous publications significantly increased the knowledge available about the region, particularly Iran. Today the most important basis for research in vegetation ecology is Rechinger's "Flora Iranica" (1963 ff.).

At present there is fairly good information available about the flora of Iran. In contrast, what is known about the sociology and ecology of numerous vegetation units, particularly in the arid parts of the country, must be called modest. Only the northern part of the country has been relatively well studied.

Descriptive studies in vegetation ecology and analyses concerning plant geography were carried out by authors such as Kotschy (1861), Melchior (1937) and Rechinger (1939). Gilli (1939) was the first to conduct studies in plant sociology. Bobek carried out a variety of geographical investigations in Iran. Among them was his fundamental study on the natural forests and woodlands of Iran (Bobek 1951). The first comprehensive works to take woody-plant formations as well as other vegetation units into account were written by Zohary (1963) and Mobayen/Tregubov (1970). In his "Geobotanical Foundations of the Middle East" Zohary (1973) dealt again in detail with the vegetation of Iran. Between 1970 and 1978 geobotanical research in the country was intensified. Important par-

ticipants in this were the Tehran Botanical Garden, the Department of Environment, Tehran, the Botany Division of the Special Research Division 19, "Tübingen Atlas of the Middle East" (Tübingen, FRG), and the Biological Department of the University of Shiraz.
A list of geobotanical literature up to 1970 is given in FREY/MAYER (1971), and a complete list up to 1984 may be found in BIERKAMP/FREY/KÜRSCHNER (in press).

3. Climate (fig. 1)

Only the south Caspian coastal plain and the northern slopes of the Elburz Mountains up to an altitude of 2500 m belong to the humid areas. The main ridge of the Elburz separates the humid Caspian region from the arid Iranian highlands. At high altitudes of the Elburz Mountains and in the mountains of Kurdistan, the Zagros range and northern Ḫorāsān,

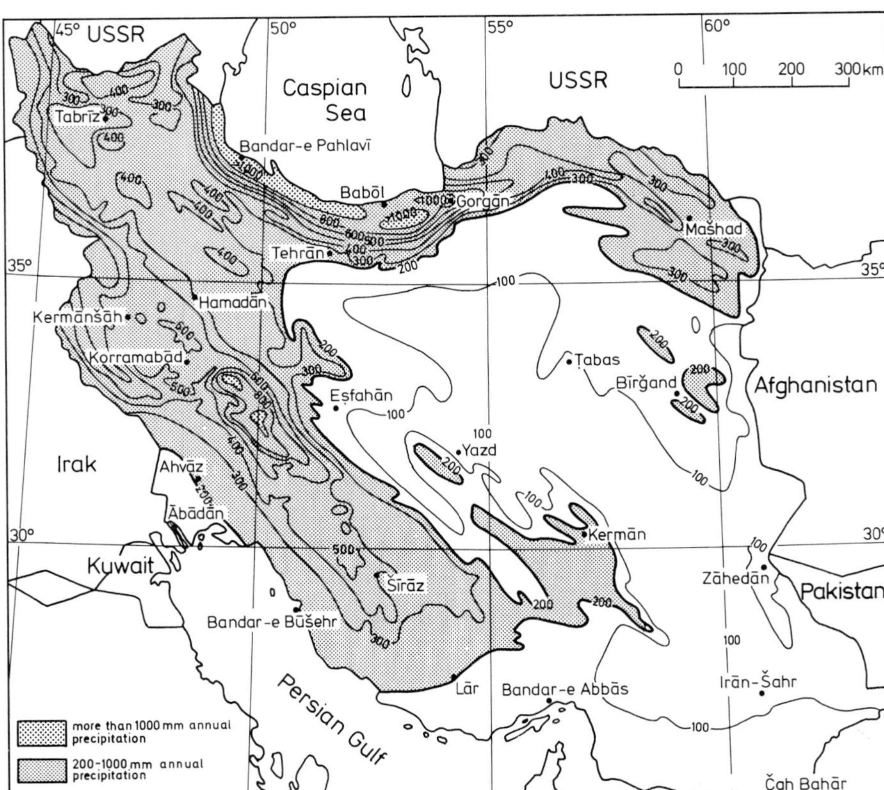

Fig. 1 Average annual precipitation in Iran (after SCHWEIZER 1975).

winter and spring precipitation is relatively high (400–800 mm/year, Elburz more than 1000 mm/year). However, the vegetation is strongly affected by the extreme summer aridity, which is interrupted only by occasional showers. Large areas of the country receive less than 200 mm of precipitation per year. This is the case for central Iran, the basin zone south-east of Eṣfahān, the interior of Baluchistan and Sistan. The coastal areas on the Persian Gulf are also very arid. In these areas the little precipitation occurs mainly in winter and spring. It is brought by low-pressure cells coming from the west.
In the south Caspian coastal lowlands the winters are especially mild. The winter minima are seldom below –5 °C. On the other hand, in north-western Iran, the Zagros Mountains and the adjoining basin regions in the Iranian interior the winters are severe. The winter minima, which are of decisive importance for plants, are often below –25 °C. The desert areas in the south and south-east of the country are subtropical hot deserts. Even so, there are night frosts here, as well as occasional snowfalls. In the area of the gulf coast Iran borders the tropical zone.

4. Phytogeographical regions of Iran
(after Zohary 1973, modified)

Most of Iran belongs to the Holarctic kingdom; the coastal areas on the Persian Gulf and the foothills there belong to the Paleotropic kingdom. The border between the two kingdoms runs through southern Iran. Since species of the Nubo-Sindian element penetrate from the south through the valley systems, it cannot be determined exactly. The transition zone is probably 100–150 km wide (Fig. 2).
The following description is based on that by Zohary (1973). There are divergent conceptions regarding the paleotropic phytogeographic units; these will be discussed in sections 4.1.4 and 4.2. On the phytogeographical regions of Iran compare also Hedge/Wendelbo (1978).

4.1 Holarctic phytogeographical units

4.1.1 The Hyrcanian district of the Caucasian-Euxino-Hyrcanian province

The Euro-Siberian region extends into northern Iran in the Caucasian-Euxino-Hyrcanian province. This Hyrcanian district includes the south Caspian coastal plain and the northern slopes of the Elburz Mountains up to 2500 m. The western and northern boundaries are formed by the Arāk, and the eastern boundary runs west of Boǧnūrd. There are a number of woody plants that are peculiar to this district, for example *Parrotia persica*, *Acer velutinum*, *Alnus subcordata*, *Amygdalus spinosissima* ssp. *turcomanica*,

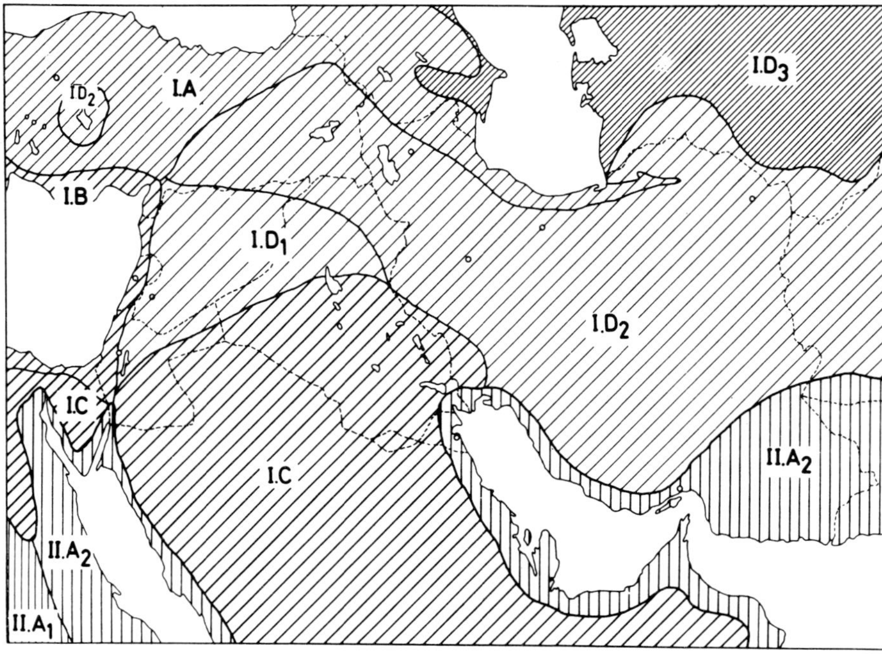

Fig. 2 Phytogeographical regions of Southwest Asia (after ZOHARY 1973, altered).
Holarctic phytogeographical units
I.A Caucasian-Euxino-Hyrcanian province of the Euro-Siberian region
I.B Mediterranean region
I.C East Saharo-Arabian subregion of the Saharo-Arabian region
I.D_1 Mesopotamian province of the Irano-Turanian region
I.D_2 Irano-Anatolian province of the Irano-Turanian region
I.D_3 Turanian province (Medio-Asiatic province)
Paleotropic phytogeographical units
II.A_1 Sahelo-Sudanian province of the Sudanian region
II.A_2 Nubo-Sindian province of the Sudanian region

Acer monspessulanum ssp. *turcomanicum*, *Quercus castaneifolia*, *Q. macranthera*, *Lonicera floribunda*, *Pyrus boissieriana*, *Sorbus orientalis*, *Ilex spinigera*, *Ruscus hyrcanus*, *Gleditsia caspica*, *Buxus hyrcana* and *Hedera pastuchovii*. There are no species of *Abies*, *Picea* or *Pinus*.
The Hyrcanian district is characterized by *Predominantly deciduous closed forests* and, in the border, by *Cold-deciduous scrub*. It contains relics of the Arcto-Tertiary and Indo-Malesian flora which retreated to the south during the ice ages and survived there.

4.1.2 The Irano-Anatolian province of the Irano-Turanian region

The Irano-Turanian region extends from central Asia west through southwest Asia and reaches as far as the Mauritanian steppe in northern Africa. The Irano-Anatolian province includes central and east Anatolia, large parts of Iran (about two-thirds of its area) and Afghanistan, and the Pakistani sector of Baluchistan. Like the whole phytogeographical region, it is characterized by a continental climate. Hot, dry summers and cold winters predominate. Precipitation occurs in the winter and in the early months of spring; it reaches high levels (up to 800 [1000] mm/year) only at high elevations in the mountains.

Typical vegetation units are *Dwarf-scrub* (highlands), *Salt swamps* (Iranian interior basins), *Woodlands* and *Thorn-cushion formations* (mountain ranges).

This province is characterized by a large number of endemic species and genera. Examples are the genera *Amygdalus*, *Astragalus*, *Pistacia*, *Ferula*, *Nepeta*, *Phlomis*, *Verbascum* (centre of endemism in Anatolia), *Acanthophyllum*, *Dionysia*, *Acantholimon*, *Cousinia* and *Eremurus*, each of which has many species. This is one of the most pronounced centres of speciation and distribution in the Holarctic kingdom; especially the montane and alpine zones of the mountain ranges exhibit extraordinary diversity. Hedge/Wendelbo (1978) mention five patterns for the montane-alpine zone, each of which has a large number of endemites: the Elburz pattern, the Zagros pattern, the Armeno-Kurdic pattern, the Khorassan pattern and the Pan-mountain pattern.

4.1.3 The Turanian province (Medio-Asiatic province) of the Irano-Turanian region

The Turanian province includes the extensive areas of desert and steppe between the Caspian Sea and the Aral Sea along with the western and south-western foothills of the central Asian mountains. It touches the north-western part of Iran (Ḫorāsān). However, there are numerous small areas within the Irano-Anatolian province that have a high percentage of species from the Turanian element. This is the case, for example, on saline soils and on dunes in the interior basins of Iran. This element comprises species of the genera *Ephedra*, *Calligonum*, *Salsola*, *Haloxylon* and *Stipa*.

4.1.4 East Saharo-Arabian subregion of the Saharo-Arabian region

According to the opinion of several authors (cf. Davis/Hedge 1971, Hedge/Wendelbo 1978) there are numerous distribution patterns suggesting that there is a phytogeographical region extending from the Sahara through Egypt, Sinai, southern Mesopotamia and Arabia to southern Iran and Sind (Saharo-Sindian in the sense of Eig, compare 4.2). However, we share the view of Zohary (1973) according to which the southernmost areas of Iran and Pakistan and the western part of India belong to the Paleotropic kingdom (compare 4.2). This conclusion is based on physiognomical and ecological analyses as well as considerations related to floristics and floral history.

According to ZOHARY's (1973) subdivision Iran is touched by the Saharo-Arabian region only in the provinces of Ilām-Poštkūh and Ḫūzestān. It must be noted that the flora of this area contains numerous species from the Irano-Turanian and Nubo-Sindian elements. This region receives less than 300 mm of precipitation per year, falling in the winter months. The winters are short and mild, the summers long, dry and hot.
Examples of species belonging to this element in Iran are *Gymnocarpos decandrum*, *Moricandia sinaica*, *Diplotaxis harra*, *Anastatica hierochuntica*, *Fagonia bruguieri*, *Lasiopogon muscoides*, *Gymnarrhena micrantha* and *Centaurea pseudosinaica*.

4.2 Paleotropic phytogeographical units

ZOHARY (1973) recognized only the Sudanian region (Nubo-Sindian province) in Iran. According to DAVIS/HEDGE (1971) and HEDGE/WENDELBO (1978) there are a number of good reasons for using the older term of EIG of a Saharo-Sindian region, although for the latter authors there is no clear differentiation between the two distributional types occurring there: the pan-African – south Arabian element and the element which is determined by those taxa which stretch from north-west Africa through Arabia to south Persia, Pakistan, and south and south-east Afghanistan.
Various investigations, concerning the physiognomy and ecology of the vegetation, migration paths of tropical species, and distribution patterns in the area occupied by species of the Nubo-Sindian element (unpublished), have shown that the Saharo-Sindian distribution patterns are not of great importance. What clearly emerges is a Nubo-Sindian province into which Saharo-Arabian and Mediterranean species have immigrated.

4.2.1 Nubo-Sindian province of the Sudanian region

There is disagreement about the dividing line between the Sudanian region and the Saharo-Arabian (Saharo-Sindian) region (see above). If one adopts the demarcation made by ZOHARY, then the northern limit of the Nubo-Sindian province in Iran probably coincides with the northern limit of *Ziziphus spina-christi* and the *Acacia* species. On account of their sensitivity to cold the species of the Nubo-Sindian element are confined to the low-lying regions on the Persian Gulf. However, numerous species are able to penetrate along the valleys between the Zagros foothill ranges. In Sistan and Iranian Baluchistan the Nubo-Sindian province stretches away from the gulf into the interior, reaching as far as south-west Afghanistan.
The flora of the Nubo-Sindian province is xerophytic and very thermophilous. It requires very high summer temperatures, high humidity and frost-free winters. *Xeromorphic, very open scrub (semidesert shrublands)* is

extremely common. Some of the characteristic species of the Nubo-Sindian province are *Capparis decidua*, *Acacia seyal*, *A. ehrenbergiana*, *Cleome brachycarpa*, *Tephrosia persica*, *Tribulus longipetalus*, *Euphorbia larica*, *Salvadora persica*, *Ziziphus spina-christi*, *Stocksia brahuica*, *Periploca aphylla*, *Calotropis procera*, *Caralluma edulis*, *Pergularia tomentosa*, *Convolvulus acanthocladus*, *C. virgatus*, *Heliotropicum laricum* and *Pennisetum divisum*.

5. The vegetation of Iran

5.1 General

The following general outline of vegetation conditions in Iran is based on a classification of vegetation according to physiognomical and ecological criteria (Frey/Probst 1977, proposed classification scheme for vegetation in south-west Asia). In the descriptions of vegetation units there are references to further reading and to literature on special topics.

5.2 Synopsis of the vegetation units

5.2.1 Closed forests

5.2.2.1 *Predominantly deciduous closed forests* (Hyrcanian forest) (photograph 1)

The *Predominantly deciduous closed forests* are restricted to the south Caspian lowlands and to the adjoining northern slopes of the Elburz Mountains up to 2500 m. The forest area extends from Ṭālеš in the west to the eastern boundary of the Golestan Park east of Gorgān. Islandlike patches may be found as far as Mehmānak, west of Boǧnūrd.

Today the lowest zone of this Caspian "Dschängall" has been replaced largely by cultivated land. The main crops are citrus fruits, tea, rice, cotton and wheat.

The forest area comprises most of the Hyrcanian district. In the area of transition to the arid parts of the Elburz Mountains the forest merges into scrub formations (compare 5.2.3). The forests contain relics of the Arcto-Tertiary and Indo-Malesian flora. The south Caspian region is apparently an ice-age refugium in which a species-poor "Pliocene forest" was able to persist. This complex was described in detail by Probst (1981).

The forest area is divided into a *Cold-deciduous lowland forest with evergreen broad-leaved trees or shrubs*, a *Cold-deciduous lowland forest* and a *Cold-deciduous montane forest.* This division is based on the distribution of evergreen broad-leaved trees and shrubs and on the upper distribution limits of a few frost-sensitive species such as *Pterocarya fraxinifolia*, *Diospyros lotus*, *Albizia julibrissin* and *Parrotia persica*. *Cold-deciduous alluvial forests*, which today are still found in a few places, cannot be discussed here.

Cold-deciduous lowland forest with evergreen broad-leaved trees or shrubs (Hyrcanian lowland forest)

The area of the southern Caspian coastal plain between –20 m and 500 m and the slopes of the Elburz Mountains west of Babol was originally covered by a *Cold-deciduous lowland forest.* The undergrowth contained numerous evergreen species and lianas (Querco-Buxetum Mobayen/Tregubov 1970). Today large areas of the plain have been cleared and are under cultivation. Although the remaining portions have been severely disturbed by man, the characteristic storey persists. Precipitation is from 1000 to 2000 mm/year, and there is high diurnal and seasonal humidity; winter minima are rarely and only briefly less than 0° C and never less than –10° C. There is an abundance of epiphytic vegetation, the characteristic species being *Palamocladium euchloron* (Musci), a paleotropical relic element. Characteristic species of the tree and shrub layer are: *Acer velutinum, Carpinus betulus, C. orientalis, Pterocarya fraxinifolia, Quercus castaneifolia, Tilia rubra, Zelkova carpinifolia*; *Albizia julibrissin, Buxus hyrcana, Diospyros lotus, Gleditsia caspica, Ilex spinigera, Parrotia persica, Prunus laurocerasus, Ruscus hyrcanus.* Lianas: *Hedera pastuchovii, Smilax excelsa, Tamus communis.*

Cold-deciduous lowland forest (Hyrcanian lowland forest)

This forest type is distributed between 500 and 700 (1000) m, east of Babol also in lower areas. In this forest type no evergreens occur (Parrotio-Carpinetum Mobayen/Tregubov 1970). In the tree layer this formation is distinguished by the absence of *Albizia julibrissin, Parrotia persica* and *Pterocarya fraxinifolia.* For the characteristic species see 2.12 (without evergreens). The upper distribution limits for some characteristic species are: *Pterocarya fraxinifolia* 500 (700) m, *Albizia julibrissin* 700 m, *Diospyros lotus* and *Gleditsia caspica* 800 m, *Ficus carica, Tamus communis* and *Vitis* spec. 1000 m. *Parrotia persica* also occurs above 1000 m, but it is dominant below 600 m.

Cold-deciduous montane forest (Hyrcanian montane forest)

At an altitude of 700–1000 m the lowland forest merges gradually into the Hyrcanian montane forest. There *Fagus orientalis* is dominant (Fagetea hyrcanica Zohary 1963, Fagetum hyrcanum, Carpinetum orientalis Mobayen/Tregubov 1970). The composition of the Hyrcanian montane forest is variable. On places with high humidity *Fagus orientalis* is dominant; on drier places *Carpinus orientalis* and *C. betulus* are the main species. In narrow valleys, at the level where Caspian air masses condense, there occurs a variant of the "cloud forest" (perhumid conditions) with species that prefer more moisture and with luxuriant epiphytic vegetation (characteristic species: *Leucodon immersus* [Musci]).

The upper limit of the montane forest is caused by dryness. In Kūhhā-ye Ṭāleš it lies between 2000 and 2100 m, in the central Elburz it is at 2400 m, and in the eastern Elburz it is near 1900 m. Above the condensation level the closed montane forest blends into woodland (see 5.2.2.3, *Cold-deciduous broad leaved montane woodland*) and scrub (see 5.2.3) owing to aridity and intensive summer grazing. Characteristic species of the tree layer are: *Acer hyrcanum*, *A. platanoides*, *Carpinus betulus*, *C. orientalis*, *Fagus orientalis*, *Quercus castaneifolia*, *Q. macranthera*, *Sorbus orientalis*, *Tilia platyphyllos* ssp. *caucasica* and *Ulmus glabra*.

The forest formations of the Hyrcanian district are the best-known vegetation units in Iran. They have been described in comprehensive studies by Mobayen/Tregubov (1970) and Zohary (1973), and have been the subject of other detailed investigations in the past decade. Probst (1972, 1974) and Frey/Probst (1974 a) supplied vegetation profiles correlated with climatic data; Mossadegh (1972) presented a survey of the Caspian forests; Asli/Nedialkov (1974) and especially Rastin (1983) carried out pioneering sociological analyses on forest units; Dorostkar (1974) investigated the forest areas near Gorgān. Frey (1980, 1983) reported on the eastern limits of the Caspian forest area and the transition to the *Dwarf-scrub formations* of the Iranian interior plateau. A vegetation map of the central part of the Elburz Mountains was prepared by Frey et al. (1985). The bryophyte vegetation of the Hyrcanian forest area is discussed in the studies by Frey/Kürschner (1977, 1979); it is shown that there is a close relation between the epiphytic vegetation and climatic conditions in the region.

Further detailed phytosociological analyses are urgently needed in order to subdivide the Hyrcanian forest formations that have been distinguished on the basis of physiognomical-ecological and floristic data.

5.2.1.2 *Mangrove forests* (photograph 2)

The *Mangrove forests* between the island of Ǧazīre-ye Qešm and the Persian Gulf coast near Bandar-e Ḫamīr (Haragheshm Protected Region) today are the most extensive and best preserved stands on the coast of the Persian Gulf. Like the other mangrove stands in this region they consist only of one species, *Avicennia marina* (Avicennietea marinae, Zohary 1973). *Rhizophora mucronata*, mentioned in the literature, is not present. The mangroves extend as far north as Bandar-e Būšehr (see vegetation map of Mobayen/Tregubov 1970). Between Ǧazīre-ye Qešm and the mainland the mangrove stands colonize mud banks. Especially in the coastal region they are separated by a vegetation-free strip from the halophyte communities (i. e. *Halocnemum strobilaceum* stands) and other species situated further inland. This zonation is typical of mangroves in arid regions (heavy salt accumulation through evaporation in the marginal coastal zone only briefly inundated at each high tide).

There are only three studies on the mangroves of Iran. DJAZIREI (1961) and KUNKEL (1977) give brief explanations on the mangroves. In a vegetation map (FREY/PROBST 1986) the distribution of the mangrove stands and the species composition of the surrounding vegetation units between Ǧazīre-ye Qešm and the mainland is shown.

5.2.2 Woodlands

BOBEK (1951) and ZOHARY (1963, 1973) have provided detailed descriptions of the floristic composition and distribution of the extensive woodlands in the mountains of Iran. Further studies have been written by FREY (1980, 1983) and FREY et al. (1985) on northern Iran (Elburz Mountains and the adjacent ranges of Ḫorāsān), by TABATABAI/JAVANSHIR (1966) and KRAMER (1984) on the Zāgros Mountains, and by CARLE/FREY (1977), FREY (1982), PONTECORVO/BOKHARI (1975) and EDMONDSON et al. (1980) on the mountain ranges between Šīrāz and Kermān. It must be emphasized that there is still no modern study based on phytosociological analyses. Ecological investigations are lacking as well.

5.2.2.1 *Evergreen needle-leaved woodland sensitive to cold (Cupressus sempervirens, Thuja orientalis)* (photograph 3)

Cupressus sempervirens and *Thuja orientalis* woodlands are known from the intramontane basins of the northern slope of the Elburz Mountains. *Cupressus sempervirens* stands are also known from Fīrūz-Ābād (southern Iran) and from the Kūh-e Taftān (south-eastern Iran) (BOKHARI 1975). The best-known locality of this forest formation is the intramontane basin of Ḥasanabād south of Čālūs. Annual precipitation is less than 500 mm. There is a retreat of deciduous forest in favour of a xerophytic, Mediterranean *Cupressus* woodland. East of Gorgān *Thuja orientalis* dominates in such dry valleys. Probably this forest type is a relic from drier climate periods of the Tertiary (Miocene) (ZOHARY 1963, PROBST 1981).

5.2.2.2 *Evergreen needle-leaved woodland resistant to cold* (photograph 4)

Large stands of open needle-leaved forest with the characteristic species *Juniperus excelsa* (Juniperetum excelsae ZOHARY 1973) are found on the southern slopes of the Elburz-Mountains north of Šāh-Rūd and in the area east of Golestan National Park (east of Gorgān) as far as the Boǧnūrd area (compare FREY 1980, 1983). The ranges near Caman-e Bīd (Kūh-e Kūrḫūd) are especially notable for extensive stands of *Juniperus excelsa* woodlands, which interlock with the Hyrcanian forest in the west. Sparse stands of *Juniperus excelsa* are scattered throughout the southern Elburz slopes (FREY et al. 1985). In the Zāgros Mountains there are large

stands only on Kūh-e Denā (Pontecorvo/Bokhari 1975). The other occurrences in the Zāgros ranges can hardly be called woodlands. *Juniperus excelsa* is also found in the mountains south of Kermān (Edmondson et al. 1980), and the southernmost outposts are in the region of the summit of Kūh-e Genū near Bandar-e Abbās.
In the area of transition to the Hyrcanian forest the *Juniperus excelsa* woodlands are heavily interspersed with deciduous shrublike species. The number of shrub species declines rapidly with the decrease in moisture as one goes east. Dwarf-shrubs and thorn-cushions take their place. The shrubs that penetrate furthest east in the *Juniperus* woodlands of Ḫorāsān are *Celtis caucasica*, *Acer monspessulanum* ssp. *turcomanicum*, *Cotoneaster nummularioides*, *Cerasus pseudoprostrata* and *Amygdalus spinosissima* ssp. *turcomanica*.
In the Zāgros ranges *Juniperus excelsa* determines the physiognomy of the woodlands and scrubs only in rare cases (see above). In section 5.2.3.3 the species making up the other units in the Zāgros are named and the reasons are discussed why these units cannot be considered related to the *Juniperus* woodlands.
In the Elburz Mountains the upper limit of the *Juniperus excelsa* woodlands is at 2500 m; the lower and eastern limits of distribution are a result of aridity. For this formation to develop a precipitation between 400 and 500 mm per year is necessary. Of this, small amounts must occur in the summer, since rejuvenation of the woodlands apparently is otherwise impossible. The climate is markedly continental. Typically there are severe winter frosts (below –25 °C), almost unmitigated summer heat, and low humidity.

5.2.2.3 *Cold-deciduous broad-leaved woodlands without evergreens* (photograph 5)

At locations with unfavourable exposure the originally closed Hyrcanian montane forest changes to woodland. The physiognomy of this formation is determined by the characteristic species *Quercus macranthera* (Quercetum macrantherae Zohary 1963), which forms loose groups between 2200 and 2600 m on the northern slopes of the Elburz Mountains. Because of the increasing aridity and the greater amount of incident light, the herb layer is dominated by taxa which are more xerophilous.
On the south-western slopes and in the central parts of the Kurdestan and Zāgros ranges, *Quercus* woodlands characterized by *Quercus brantii* and related species predominate (Quercetum brantii Zohary 1973). Their south-eastern limit of distribution is south-west of Šīrāz. On the side towards the Mesopotamian lowlands they are bordered by a *Pistacia* woodland (*Pistacia khinjuk, Ziziphus nummularia*; compare Kramer 1984) and by a belt of shrubs.

Their species diversity declines from north-west to south-east, corresponding to the decrease in precipitation. In the direction of the Iranian interior highland the woodlands are succeeded by a drier type with a high percentage of *Juniperus excelsa*, followed by *Astragalus* scrub and finally by the *Dwarf-scrub formations* of the highland.

Quercus species are the dominant trees. In the north-west *Quercus brantii* is joined by Q. *infectoria*, Q. *boisseri* and Q. *iberica*. In the central Zāgros only Q. *brantii* occurs. As the species best adapted to drought it is able to penetrate further south-west than the others. Accompanying tree species are *Fraxinus rotundifolia* ssp. *persica*, *Acer monspessulanum* ssp. *cinerascens*, *Pistacia khinjuk*, *Amygdalus elaeagnifolia*, *Pyrus syriaca* and *Pistacia atlantica* ssp. *mutica*. In some places *Pyrus syriaca* and *P. glabra* are dominant.

These *Quercus* woodlands occupy a band at altitudes between 1000 and 2100–2300 m. Their lower limit is caused by aridity, and above 2300 m they merge into a *Quercus* scrub formation which is then succeeded by the vegetation units of the alpine zone. Annual precipitation in the area occupied by *Quercus* woodlands in the Zāgros range is more than 400 mm; on mountain ridges it is higher, exceeding 800 mm in some cases. For these woodlands to occur it is necessary for there to be fairly regular summer precipitation, generally in the form of thundershowers. This ensures rejuvenation of the stands. The minima in the winter months are not as low as those in the *Juniperus* woodlands.

Zohary (1973) gave this *Quercus* woodland the name "Kurdo-Zagrosian oak steppe-forest" (Quercetea brantii). Bobek (1951) called it the "Eichenwald der Zagrosketten". Tabatabai/Ivanshir (1966) also describe these forests in detail.

Today large areas of these oak forests have been destroyed. In the areas that have been cleared rainfed cultivation is practised, the main crop being wheat.

5.2.2.4 *Extremely xeromorphic woodlands* (Pseudosavannas)

Pseudosavannas occur in the coastal area of the Persian Gulf only sporadically. The potential woodland areas are severely degraded and are today occupied by *Open xeromorphic scrub*. Sociologically the Pseudosavannas of Iran belong to the class Acacietea flavae iranicae (Zohary 1963). Only in a few places are there remnants which are in a relatively natural state. Frey/Probst (1986) studied a well-developed Pseudosavanna on Ğazīre-ye Qešm. Characteristic species are: *Acacia aucheri*, *A. nubica*, *Prosopis cineraria*, *Crotolaria furfuracea*, *Gymnocarpos decandrum*, *Sphaerocoma aucheri* (trees and shrubs); *Farsetia heliophila*, *Jauberta aucheri*, *Taverniera cuneifolia* (dwarf-shrubs). A second floristic analysis of a Pseudosavanna in Makran is given by Assadi/Runemark (1983). Dominant tree species are *Acacia ehrenbergiana* (tree/shrub), *Ziziphus spina-christi* and *Prosopis cineraria*.

5.2.3 Scrub

Scrub often covers large areas on the margins of forests and woodlands. To date it has been subdivided only according to physiognomical-ecological and floristic criteria. Phytosociological analyses are lacking.

5.2.3.1 *Evergreen needle-leaved scrub*

Evergreen needle-leaved scrub has been found so far only in the subalpine zone of the Elburz Mountains and in the adjacent mountain ranges of Ḫorāsān. *Juniperus excelsa* and *Juniperus communis* ssp. *communis* are dominant. A hedge-like habit of *Juniperus excelsa* occurs in high altitudes of the southern Zagros Mountains (Pontecorvo/Bokhari 1975).

5.2.3.2 *Cold-deciduous scrub* (photograph 6)

The northern Elburz slopes between 2200 and 2700 m (upper montane zone) are covered in many places by scrub which, on account of the exposure, can extend down to 1600 m where it merges into open scrub. At very high locations, depending on exposure and moisture supply, it is made up of species of the Hyrcanian montane forest or it is dominated by more xerophilous species of the *Cold-deciduous broad-leaved woodland* (Quercetum macrantherae) such as *Berberis orthobotrys* and *Colutea persica.*
At the eastern limit of the Elburz Mountains numerous species of the Hyrcanian forests are absent on account of lower precipitation. *Zelkova carpinifolia, Carpinus orientalis, C. betulus* and *Quercus macranthera* are able to penetrate farther eastward in shrub form and are components of the scrub in the mountain ranges of Ḫorāsān. Here there occur other characteristic scrub species such as *Acer monspessulanum* spp. *turcomanicum, Acer cappadocicum, A. campestre, Crataegus spp., Mespilus germanica, Prunus divaricata, Cerasus microcarpa, Pyrus boissieriana, Colutea persica* and *Celtis caucasica* (Frey 1980, 1983).
This scrub is dependent on regular summer precipitation, but drought periods of some duration can intervene. The total annual precipitation is probably about 500 mm. Low minima in the winter months are also characteristic.

5.2.3.3 *Open xeromorphic scrub* (photograph 7)

On the south-western slopes of the Zāgros ranges *Open xeromorphic scrub* accompanies the *Quercus brantii* woodlands towards the Mesopotamian lowlands and the lowlands on the Persian Gulf. It extends past Šīrāz, where the *Quercus brantii* woodlands cease, and continues into the southern and south-eastern mountain ranges and massifs. The outermost occurrences are at 27° N, east as far as Kūh-e Taftān, south of Zahedan and on Kūh-e Genū near Bandar-e Abbās. Here, however, they often merge into *Xeromorphic, very open scrub (semidesert shrublands).*

Bobek (1951) referred to the *Open xeromorphic scrub* as "Pistazien-Mandel-Ahorn-Trockenwald", Zohary (1973) called it "*Juniperus-Pistacia-Amygdalus*-steppe scrub" (Junipero-Pistacietea). Kramer (1984) divided this scrub into three variants: a) scrub with thorny *Amygdalus* species (dominant), b) *Amygdalus scoparia* scrub and c) *Lonicera nummulariifolia* scrub on Kūh-e Dinā.

Further information is given in Carle/Frey (1977), Edmondson et al. (1980) and Kunkel (1975). The characteristic species are *Amygdalus scoparia*, *A. eburnea*, *A. elaeagnifolia*, *A. kotschyii*, *A. ericlada*, *Acer monspessulanum* ssp. *cinerascens* and *A. m.* ssp. *persicum*, *Pistacia khinjuk* and *P. atlantica.* Other species are *Ephedra ciliata*, *E. pachyclada*, *E. strobilacea*, *Rosa* spp., *Cerasus microcarpa*, *Cotoneaster persica*, *Colutea gracilis*, *Daphne mucronata*, *Prunus divaricata*, *Rhamnus pallisii*, *R. kurdicus*, *Ficus carica* and *Juniperus excelsa.*

Although the *Acer* and *Pistacia* species are frequently arboraceous, they are shrubs in the physiognomical sense. In the south-east and east, where the *Pistacia* stands are often only sparse, the coverage is so low that one can no longer speak of woodland or scrub. This "Bergmandel-Pistazien-Baumflur" (Bobek 1951) (photograph 8) must be classified either as *Xeromorphic, very open scrub* or as *Dwarf scrub.*

In the north-west the scrub overlaps with the *Quercus brantii* woodlands. A large number of species are present, and it is often impossible to make a clear demarcation. As one moves south-east there is less summer and annual precipitation, and the number of species declines accordingly. At Šīrāz, where annual precipitation in the scrub region amounts to about 400 mm, a large number of species are already absent. Bokhari/Khan (1976) analysed the distribution of species on various geological substrata. In the area that was surveyed rock formations appear to determine the distribution of various shrubs.

The most drought-resistant species that penetrate furthest on the massifs of south-eastern Iran are *Pistacia atlantica*, *P. khinjuk*, *Acer monspessulanum* ssp. *persicum*, *Amygdalus haussknechtii*, *Cerasus microcarpa* and *Juniperus excelsa.*

The lower limit of the scrub is determined by the amount of precipitation received (300–350 mm/year). The upper limit of distribution is at 3000 m.

Zohary (1973) grouped the *Juniperus* woodlands and the *Pistacia-Amygdalus* scrub together as "*Juniperus-Pistacia-Amygdalus*-steppe scrub" (Junipero-Pistacietea). We do not share this view, since the ecological requirements and the species composition of these two vegetation units differ so much that they must be distinguished phytosociologically. Pure *Juniperus* woodlands are limited to areas with low minima during the winter months (–25 °C and below) and irregular summer precipitation. In contrast, the *Amygdalus-Pistacia* scrub requires higher winter temperatures and more or less regular summer precipitation.

The *Open xeromorphic scrub* also includes the *Tamarix* scrub on the saline soils of the southern and eastern interior plateau and the gulf region. In the wadis of the gulf region *Tamarix aphylla* can form large stands of trees.

5.2.3.4 *Xeromorphic, very open scrub (Semidesert shrublands)* (photographs 9, 10)

In areas with low precipitation (less than 300 mm/year) the *Open xeromorphic scrub* merges into *Xeromorphic, very open scrub.* The most drought-resistant representatives are *Amygdalus* spp., *Zygophyllum eurypterum* and *Pistacia khinjuk.* Little is known to date about their regional distribution. Moore/Bhadresa (1978) are the only investigators to have analysed a stand of *Zygophyllum eurypterum* from the point of view of plant ecology. The stand was located in the Touran Biosphere Reserve in north-eastern Iran. They noticed that the dispersion of bushes supports the hypothesis of infra-specific competition as the primary limitation to the *Zygophyllum* populations. The Touran Biosphere Reserve was also the site of a floristic analysis of *Salsola-Zygophyllum* communities (Freitag 1977). There *Zygophyllum eurypterum* is one of the dominant species.
Xeromorphic, very open scrub (semidesert shrublands) is characteristic of the lowest hill zone in the coastal regions of the Persian Gulf, especially in Makrān and Lārestān. It is found predominantly on karstic limestone formations, and typical species are *Acacia nubica* and *Euphorbia larica* (*Acacia nubica – Euphorbia larica* zone, Euphorbietea laricae Zohary 1973). Shrubs are characteristically dominant, followed by dwarf-shrubs. Hemicryptophytes are almost completely absent. After sufficient winter rains there are numerous therophytes in spring. Characteristic shrubs (beside those mentioned above) are: *Capparis cartilaginea, C. spinosa* ssp. *mucronata, Ephedra ciliata, Gymnocarpos decandrum, Lycium depressum* ssp. *angustifolium, Ochradenus baccatus, Pteropyrum olivieri, Zygophyllum eurypterum* ssp. *eurypterum.* Frey/Probst (1986) give a preliminary floristic analysis.

5.2.4 Dwarf-scrub and thorn-cushion formations (photographs 8, 11)

Dwarf-scrub in the proper sense and *Thorn-cushion formations* are the characteristic vegetation units in Iran besides *Halophytic communities. Dwarf-scrub* is typical of large areas in the interior of Iran and of the slopes of the ranges bordering the interior regions. *Thorn-cushion* formations characterize the subalpine zone of the Iranian mountains and degraded *Juniperus excelsa* woodlands (compare vegetation map of Mobayen/Tregubov 1970). It is therefore surprising that these two units, together with the *Halophytic communities,* are the ones in the Iranian interior regions about which the least is known. ZOHARY (1963, 1973) was the first to attempt a more exact classification. At present it is impossible to give a more accurate analysis, although a number of studies have now appeared containing information on distribution, sociology and floristics (BRECKLE 1983, BOKHARI/KHAN 1976, CARLE/FREY 1977, EDMONDSON et al. 1983, FREITAG 1977, FREY 1980, 1982, 1983, FREY et al. 1985, GHORBANLI/LAMBINON 1978, KRAMER 1984, RECHINGER 1977, RECHINGER/WENDELBO 1976).

5.2.4.1 Dwarf-scrub in the proper sense (photograph 8)

The extent, density and species composition of the dwarf-scrub (végétation steppique MOBAYEN/TREGUBOV 1970, Artemisietea herbae-albae iranica ZOHARY 1963) are influenced by edaphic factors, annual precipitation and duration of the dry season.
The dwarf-scrub is very rich in species and very diverse. ZOHARY (1973) distinguishes a large number of associations (i.e. *Artemisia fragrans-Atraphaxis spinosa*, *Artemisia herba-alba – Stipa hohenackeriana*, *Artemisia herba-alba – Zygophyllum atriplicoides*) depending on what dwarf-shrubs are dominant and what the accompanying species are. In a physiognomical sense one can observe a succession from herb-rich *Dwarf-shrublands* to *Xeromorphic dwarf-shrublands* to extremely open *Dwarf-shrublands.*

Mixed dwarf-scrub and herbaceous formations

Mixed dwarf-scrub and herbaceous formations are typical of many parts of the upper montane zone of the Elburz and Kurdo-Zagrosian Mountains between 2400 and 3000 m. Owing to relatively high precipitation (up to 500 mm) this is the most species-rich variant of the Irano-Turanian steppe formation. FREY et al. 1985 describe a species-rich example from the Elburz Mountains.

Xeromorphic dwarf-shrublands

On the peripheries of the Iranian mountain ranges (extending up to 1800 (2000) m herbaceous species retreat on account of low precipitation (200–250 mm/year); xerophilous dwarf-shrubs dominate (Artemisietea herbae-albae iranica ZOHARY 1963). In many places intensive utilization by man and animals has led to the formation of substitute communities (*Peganum harmala* steppe, nitrophilous substitute community, *Alhagi maurorum* community, segetal-ruderal substitute community or secondary thorn-cushion formations).

Xeromorphic dwarf-shrublands with thorn-cushions

In the foothills of the mountain ranges tragacanthic *Astragalus* and *Acantholimon* species become more common. A dwarf-scrub variant with many thorn-cushions is formed (Artemisietea herbae-albae astragaletosum glaucacanthi ZOHARY 1973). This thorn-cushion steppe is also characteristic of large parts of the nonsaline interior region of the Iranian highland.

Xeromorphic, very open dwarf-shrublands

Under extremely arid conditions the open dwarf-shrublands change into a very open variant. This formation is characteristic of great areas of the Iranian interior regions. BRECKLE (1983) describes such very open dwarf-shrublands from the Touran Biosphere Reserve (northeastern Iran); FREY et al. report on examples from the plains on the southern margin of the Elburz mountains. The dominant species are *Artemisia herba-alba, Aellenia glauca, Astragalus gossypius, Dendrostellera lessertii* and *Pteropyrum aucheri.*

5.2.4.2 *Open thorn-cushion formations* (photograph 11)

Thorn-cushion formations are one of the most characteristic vegetation units of Southwest Asia. This vegetation unit occurs originally in the subalpine zone of the mountain ranges and as a substitute unit in disturbed woodlands.

The *Open thorn-cushion formations* of the Iranian mountains are classified by ZOHARY (1973) within the Astralagetea iranica. Although they are among the most characteristic vegetation units of all, very few studies have been written about them up to now (GILLI 1939, ZOHARY 1973, FREY et al. 1985). The *Thorn-cushion formations* of Iran have not yet been the subject of a modern analysis of vegetation such as that carried out by KÜRSCHNER (1982) for the *Thorn-cushion formations* of the Aladağları and Erciyes Dağı in Turkey.

This unit covers large areas in the subalpine zone of the Elburz Mountains, the ranges of Kurdistan and the Zāgros Mountains. It is also common in the mountains of Ḫorāsān and in the upper parts of the southern Iranian massifs. Thorn-cushions are conspicuous for penetrating into degraded *Juniperus* woodlands and other disturbed vegetation units; the woodlands of Ḫorāsān are an especially good example. This suggests that *Thorn-cushion formations* below the subalpine zone are often of secondary origin and would occupy much smaller areas under natural conditions.

Characteristic species in the Elburz Mountains are *Acantholimon bodeanum, A. cephalotoides, A. demawendicum, A. dianthifolium, A. erinaceum, A. festucaceum, A. gadukense, A. hohenackeri, A. ophiocladum, A. rudbaricum, Acanthophyllum microcephalum, Arenaria gypsophiloides, Astragalus curviflorus, A. distans, A. fasciculifolius, A. glaucacanthus, A. gossypinus, A. microcystis, A. mucronifolius, A. naftolskyi, A. semipellitus, Minuartia juniperina* and *Onobrychis cornuta.*

5.2.5 Herbaceous terrestrial communities

5.2.5.1 *Grasslands*

It is probable that grasslands originally occupied more of the Iranian interior plateau than today. Although they were pushed back as a result of millennia of overgrazing, they still are of great importance in the high montane and subalpine zones of the northern Elburz slopes. FREY et al. (1985) distinguish four units there: *Grassland with trees, Grassland with thorn-cushions, Meadows and pastures.*
Owing to grazing and felling of trees, the upper montane forest zone on the northern slopes of the Elburz Mountains merges into grassland with isolated trees (between 2000 and 2400 m on 'Alam-Kūh, Gerde-Kūh, Sumanus, Lata-Kūh). *Grassland with thorn-cushions* forms a distinct unit above the tree limit in the subalpine zone, also on the northern slopes of the Elburz Mountains. *Meadows and pastures* occur on plains on the northern side of the Elburz Mountains, i.e. the plain of Kelār-Dašt and the Koğūr basin. The potential natural vegetation here is the Hyrcanian montane forest.

5.2.5.2 *Forb vegetation*

Forb formations are important especially in the subalpine and alpine zones of the mountains. They nowhere cover areas that are as large as those occupied by the previously discussed vegetation units, but they nevertheless harbour numerous endemic species from the genera *Salvia, Phlomis, Nepeta, Eryngium, Ferula, Cousinia* and others.
In the subalpine zone of the Elburz Mountains forb formations with thorn-cushions replace *Open thorn-cushion formations*, especially in the Kandūān pass area. Along melt-water streams, in avalanche tracks and erosional channels, and next to the major rivers in the Elburz Mountains there occurs a species-rich perennial forb formation with *Heracleum persicum* and *Ligularia persica* and river-gravel formations.

5.2.6 *Herbaceous and semi-woody salt swamps* (photograph 12)

Herbaceous and semi-woody salt plant formations, which characterize the extensive areas of saline soil in Iran, are still among the most poorly known vegetation units in the country.
ZOHARY (1973) classified the halophytic communities of Iran with the Halocnemetea strobilacei irano-anatolica. However, since their composition varies considerably, it can be expected that these units will be divided among several classes when they have been analysed sociologically.
The distribution of halophytic communities in Iran has been depicted cartographically by MOBAYEN/TREGUBOV (1970), ZOHARY (1963, 1973), FREI-

tag (1977, Touran Biosphere Reserve), Kramer (1984, gulf region near Bandar-e Rīg), Carle/Frey (1977) and Frey (1982) (Mahārlū basin near Šīrāz) and Frey et al. (1985, northern peripheral region of Dašt-e Kevīr). According to the presently known floristic, geographical and climatic data, four halophyte biotopes can be distinguished in Iran:

a) Salt pans (kevīrs) of the central Iranian undrained basins, and their peripheries

On account of their insufficient water supply and high summer evaporation the margins of the basins (Kevīrs) of the Iranian highland are characterized by clays and sand soils with a high salt content at the surface. Here one finds halophytic communities of varying composition. Characteristic species are: *Aellenia glauca*, *A. subaphylla*, *Aeluropus litoralis*, *Anabasis setifera*, *Cressa cretica*, *Cornulaca leucocantha*, *Frankenia hirsuta*, *Glaux maritima*, *Halocharis sulphurea*, *Halocnemum strobilaceum*, *Halotis pilosa*, *Haloxylon aphyllum*, *H. persicum*, *Salsola arbuscula*, *S. aperta*, *S. dendroides*, *S. incanescens* and *Seidlitzia rosmarinus* (Frey et al. 1985). Further floristic data on such halophytic communities are given by Ghorbanli/Lambinon (1978, Qom lake), Freitag (1977) and Breckle (1983, Touran Biosphere Reserve), Zohary (1973, northern border of Dašt-e Kevīr). The inner parts of the Kevīrs are almost free of vegetation. Physiognomically they belong to the salt desert formations.

b) Shore zones of the salt lakes

The vegetational zonation on the coast of the Mahārlū salt lake near Šīrāz is fundamentally different from that of the Kevīrs and other salt pans. The areas with high salinity can be colonized only by *Salicornia europaea* and *Halopeplis pygmaea*. In areas further from the lake these species are joined by *Spergularia marina*, *Suaeda* spec., *Cressa cretica*, *Polypogon monspeliensis*, *Aeluropus litoralis*, *Frankenia persica* and *F. pulverulenta*. Further inland dwarf-scrub begins (Carle/Frey 1977, Frey 1982, Frey/Probst 1974b).

c) Areas on the coast of the Persian Gulf

Under the influence of sea water and of salt plugs located in the coastal range, extensive halophyte swamps have formed along the gulf coast.
A typical zonation occurs in the area of the *Mangrove forests* on the coast. In the tidal zone there is a narrow band of *Avicennia marina* which is more pronounced in some regions than others. In the adjacent zone, which is inundated only at spring tide, the salt is so highly enriched that plant growth is impossible. This zone is followed by a *Halocnemum strobilaceum* community, which is replaced further from the coast by semi-woody shrubs and forbs that are less salt-resistant. In the next zone, occupied by open therophyte-rich dwarf scrub with isolated shrubs and

trees (*Acacia* spp., *Prosopis cineraria*), there is hardly any salt influence (Kunkel 1977, Frey/Probst 1986).

d) South Caspian coastal zone

The climate in the south Caspian coastal zone is very humid in the west, and precipitation declines as one moves east. At Gorgān there is already a dry period during the summer. Accordingly, the influence of salt on the soil of the coastal plain is much greater in the east (e.g., on the Gorgān Bay) than in the west. In low-lying areas one finds *Tamarix*, *Halocnemum*, *Salsola* and *Atriplex* stands; around the mouth of the Gorgān River there occur extensive halophyte swamps.

e) The zonation of the halophytes and its correlation with eco-physiological phenomena (C_3 and C_4 species distribution) was investigated by Borchers et al. (1982), Frey/Kürschner (1983), and Frey et al. (1985). Examples are the halophytic communities at the Mahārlū salt lake and near Bandar-e Ḫāmīr. It could be shown that in both examples there is a correlation between photosynthetic pathway type, soil moisture and soil salinity. The C_4 plant distribution is not only related to high summer and winter temperatures and to soil moisture, but is also influenced by salinity. The C_4 species colonize a transition zone between C_3 hygrohalophytes and xerophytic, nonhalophytic C_3 species.

5.2.7 Ephemeral forb communities

In the driest parts of the country and in ruderal areas the physiognomy of the vegetation is determined by annuals after winter and spring rains. In the *Ephemeral forb communities*, which are rich in species, the characteristic plants are from the genera *Clypeola*, *Aethionema*, *Erophila*, *Medicago*, *Trigonella*, *Cerastium*, *Spergularia*, *Spergula*, *Arenaria*, *Silene*, *Erodium*, *Papaver*, *Glaucium*, *Hypecoum*, *Roemeria*, *Filago*, *Anthemis*, *Senecio*, *Gnaphalium*, *Lactuca*, *Sonchus*, *Plantago* and others.

5.2.8 Vegetated sand dunes

The vegetation of the dunes in the interior regions, such as those near Yazd or between Bīrğand and Mašhad in Ḫorāsān, has not yet been sufficiently investigated. The following characteristic species have been reported: *Stipagrostis pennata*, *Calligonum stenopterum*, *C. bungei*, *Heliotropium* spp., *Carex physodes* and *Ephedra strobilacea* (Zohary 1963).
Frey/Probst (1974c) published a transect through the coastal dunes on the Caspian Sea near Babolsar, and Frey et al. (1985) carried out a physiognomical-ecological classification. There are especially wide strips of

dunes between Mahmūdābād and Meyan Molk. They exhibit the typical structure ranging from mobile, almost vegetationless primary dunes via secondary dunes to consolidated, densely overgrown tertiary dunes. The individual communities, which have a pronounced Mediterranean character, are dependent in their development on the exposure, moisture, salinity, mobility and texture of the dunes.

5.2.9 Deserts

Here we follow the definition given by Zohary (1973) and use the term "desert" for areas in which the direct precipitation is insufficient for plant growth.

The major deserts of Iran formed in the two large undrained basins of the interior plateau. Dašt-e Kevīr, in the north, is a vast salt desert which is free of vegetation in the centre; Dašt-e Lūṭ, in the south, is one of the hottest deserts of the world.

Breckle (1983) describes the north-eastern periphery of Dašt-e Kevīr. Isolated *Halocnemum strobilaceum* and *Nitraria schoberi* shrubs penetrate into the margin of the vegetation-free areas. In Dašt-e Lūṭ the huge plantless zones are bordered by *Tamarix aphylla*, *T. macrocarpa*, *Haloxylon ammodendron*, *Prosopis cineraria*, *Calligonum* spec. and *Seidlitzia rosmarinus* (Mobayen 1975, 1976).

Dašt-e Lūṭ has become well known from the floristic point of view through the publications of Leonard (1981–1984). No data on vegetation ecology are known besides those reported by Breckle (1983), Freitag (1977) and Mobayen (1975).

5.2.10 Plant formations of the alpine region (photograph 13)

The plant formations of the alpine regions of the Elburz and Zāgros Mountains are poorly known at present. The only phytosociological investigations to date are those that were carried out in the Elburz Mountains by Gilli (1939), Frey et al. (1985) and Klein (1982). The map of the central Zāgros Mountains published by Kramer (1984) contains no data on vegetation units.

Klein (1982) describes the chionophilous vegetation units of the central Elburz Mountains. He distinguishes five phytosociological units: Oxytropidetea persicae, Trachydietalia depressae, Catabroselletalia parviflorae, Jurinelletum frigidae and Erigeronetum elbursensis. Frey et al. discuss the *High mountain scree formation*. This formation is characteristic of large areas of the alpine zone (3600–4000 [4500] m) in the Elburz Mountains. Characteristic species are *Alyssopsis mollis*, *Androsace villosa*, *Arabis caucasica*, *Astragalus heterochrous*, *Campanula lourica*, *Centaurea zuvandica*, *Chor-*

ispora tenella, *Dielsiocharis kotschyi*, *Didymophysa aucheri*, *Draba aucheri*, *Dracocephalum aucheri*, *Erysimum repandum*, *Jurinella frigida*, *Pedicularis caucasica*, *Saxifraga iranica*, *Vavilovia formosa*, *Veronica aucheri* and *V. kurdica.*
No investigations are known concerning the ecology of alpine plant formations and the very pronounced exposure-induced variations that occur in this zone.
In the Elburz Mountains the Nival region is above 4000 m, where summers are dry and solar radiation is intensive. In northern exposures some snowfields extend down to altitudes as low as 3300 m. Large areas of glaciation exist only in the ranges of ʿAlam-Kūh and Kūh-e Damāvand.

6. Man and vegetation

With the beginning of crop farming and cattle raising mankind entered a new phase of its development. Since then man has exerted a steadily increasing influence on nature.
Farming necessarily causes changes in the vegetation, and often it brings about an irreversible alteration of the natural ecosystems through leaching of the soil and erosion. In the south Caspian lowlands and in the areas of rainfed cultivation in the mountains of Ḫorāsān and the Zāgros ranges, the natural vegetation units have been reduced to mere remnants as a result of agricultural activity.
Overgrazing of woodlands, scrub and formations free of woody plants has led to changes in the vegetation. There has been not only a reduction in coverage, but a rigorous selection of species, resulting in a restructuring of the species composition in the course of the past millennia.
The large distribution of vegetation units dominated by thorny shrubs, thorny dwarf-shrubs and thorn-cushions is certainly a consequence of overgrazing as well. The original species were pushed back, and an "ecological" niche was created in which thorny species could establish themselves and develop.

Destruction and alteration of woody-plant formations (photograph 14)

Large free-standing trees (shade trees) and sacred groves can be found in numerous locations in today treeless areas of the Iranian interior regions. These woody plants are remnants of past vegetation and enable conclusions to be drawn about its extent.
Since the thin stands of trees on the margins of natural tree habitats were presumably the preferred areas of settlement in south-west Asia, it is probable that the vegetation of these areas has changed the most. There is no doubt that the woodlands and scrubs have been pushed back severely.

However, degradation has also reached an advanced stage in densely forested areas of the Zāgros Mountains (frequently promoted by karstification) and in the humid Hyrcanian forests. The species composition of these forests has changed, because species that are better able to withstand grazing and pruning have a clear selection advantage. Many species, such as *Carpinus betulus*, *C. orientalis*, *Quercus macranthera*, *Juniperus excelsa* and *Quercus brantii*, today occur over wide areas only as low shrubs.

The influence of man on the vegetation in Iran can be summarized as follows:

In many marginal tree habitats woody-plant formations were destroyed and replaced by *Thorn-cushion formations* or *Ephemeral forb communities* (southern Elburz slopes, Zāgros border ranges, large areas of Fārs Province, climatically unfavourable slopes of the central Iranian massifs).

In the degraded forest areas there occur secondary communities containing mostly thorn-shrubs and thorn-cushions of the genera *Crataegus*, *Prunus*, *Cerasus*, *Amygdalus*, *Rhamnus*, *Paliurus*, *Astragalus*, *Ebenus*, *Acantholimon*, *Acanthophyllum* and *Convolvulus*. These communities are therefore almost useless from the economic point of view.

The species composition of the plant formations without trees and shrubs has changed as well. An example is the *Artemisia herba-alba* community, which in north-eastern Iran has largely been replaced by a *Peganum harmala* community. In large parts of the Iranian interior plateau grasses have been driven out of the dwarf-scrub formations and replaced by thorn-cushions.

In conclusion, however, it must be emphasized that the degradation of vegetation in Iran is still by no means as extensive as in other countries of south-west Asia.

7. References

ASLI, A./NEDIALKOV, S., Recherches sur la structure des peuplements vierges de hêtre oriental (*Fagus orientalis* LIPSKY) dans les futaies du nord de l'Iran, Meded. Fac. Landbouwwet Rijksuniv. Gent *39*, 1974, 78–92.

ASSADI, M./RUNEMARK, H., Notes on the flora and vegetation of S. Baluchistan, Iran, Iran. Journ. Bot. *2*, 1983, 69–78.

BIERKAMP, M./FREY, W./KÜRSCHNER, H., Bibliography of the geobotanical literature on Southwest Asia, Beih. Tübinger Atlas Vorderer Orient, Reihe A (Naturwiss.) *25*, Wiesbaden (in press).

BOBEK, H., Die natürlichen Wälder und Gehölzfluren Irans, Bonner Geogr. Abh. *8*, 1951, 62 pp.

BOKHARI, M.H., A new record of *Cupressus* from SW Iran, Not. R.B.G. Edinb. *33*, 1975, 445–447.

BOKHARI, M. H. / KHAN, R. A., Preliminary observations on plant-rock type relationships in the Shiraz-Dashte Arjan area, Southwest Iran, Vegetatio *31*, 1976, 193–199.

BORCHERS, E. / FREY, W. / HILGER, H. H., C_4 species in the highlands of Iran and Afghanistan, Not. R. B. G. Edinb. *40*, 1982, 99–114.

BRECKLE, S.-W., Temperate deserts and semi-deserts of Afghanistan and Iran. In: WEST, N. E. (Ed.), Temperate deserts and semideserts, Ecosystems of the world *5*, 1983, 271–319.

CARLE, R. / FREY, W., Die Vegetation des Mahārlū-Beckens bei Šīrāz (Iran) unter besonderer Berücksichtigung der Vegetation im Bereich der Süß- und Salzwasserquellen am Seeufer, Beih. Tübinger Atlas Vorderer Orient, Reihe A (Naturwiss.) *2*, 1977, Wiesbaden.

DAVIS, P. H. / HEDGE, I. C., Floristic links between N. W. Africa and S. W. Asia, Ann. Naturhist. Mus. Wien *75*, 1971, 43–57.

DJAZIREI, M. H., Les peuplements d'*Avicennia* des côtes suds de l'Iran, Salinity problems in the arid zones, Arid Zone Research *14*, UNESCO, 1961, 139–140.

DOROSTKAR, H., Contribution à l'étude des forêts du district hyrcanien oriental (Chaîne du Gorgan), 2 vols., 1974, Gembloux (Fac. Sci. Agronomique de l'Etat).

EDMONDSON, J. R. / MILLER, A. G. / PARRIS, B. S., Plants of the Khabr va Ruchoun Protected Area, S-Iran, Not. R. B. G. Edinb. *38*, 1980, 111–124.

FREITAG, H., Touran Biosphere Reserve. Preliminary vegetation map. In: SPOONER, B. (Ed.), Case study on desertification. Iran: Turan, 1977, 86–89, Teheran.

FREY, W.: Wald- und Gebüschverbreitung in Nordwest-Ḫorāsān (Nordiran), Beih. Tübinger Atlas Vorderer Orient, Reihe A (Naturwiss.) *6*, 1980, Wiesbaden.

FREY, W., Mahārlū-Becken bei Shiraz/Šīrāz (Iran). Mittlerer Teil. Vegetation. 1:100 000, Karte A VI 10.2 des Tübinger Atlas des Vorderen Orients, 1982, Wiesbaden.

FREY, W., Elburz (Iran). Ostgrenze des kaspischen Waldgebietes. Vegetation. 1:100 000, Karte A VI 10.5 des Tübinger Atlas des Vorderen Orients, 1983, Wiesbaden.

FREY, W. / KÜRSCHNER, H., Studies on the bryophyte flora and vegetation of the Mohammad Reza Shah National Park, N. Iran, Iran. Journ. Bot. *1*, 1977, 137–153.

FREY, W. / KÜRSCHNER, H., Die epiphytische Moosvegetation im hyrkanischen Waldgebiet (Nordiran), Beih. Tübinger Atlas Vorderer Orient, Reihe A (Naturwiss.) *5*, 1979, Wiesbaden.

FREY, W. / KÜRSCHNER, H., Photosyntheseweg und Zonierung von Halophyten an Salzseen in der Türkei, in Jordanien und im Iran, Flora *173*, 1983, 293–310.

FREY, W. / KÜRSCHNER, H. / PROBST, W., Südkaspisches Tiefland und Elburzgebirge (Iran). Vegetation. 1:500 000, Karte A VI 5 des Tübinger Atlas des Vorderen Orients, 1985, Wiesbaden.

FREY, W. / KÜRSCHNER, H. / STICHLER, W., Photosynthetic pathways and ecological distribution of halophytes from four littoral salt marshes (Egypt/Sinai, Saudi Arabia, Oman and Iran), Flora *177*, 1985, 107–130.

FREY, W. / MAYER, H.J., Botanische Literatur über den Iran, Bot. Jb. Syst. *91*, 1971, 348–382.

Frey, W./Probst, W., Vegetation und Klima des Zentralelburs und der südkaspischen Küstenebene (Nordiran), Marburger geogr. Schriften *62*, 1974 a, 93–116.

Frey, W./Probst, W., Vegetationszonierung an Süßwasserquellen des Maharlu-Salzsees bei Shiraz (Iran), Vegetatio *29*, 1974 b, 109–114.

Frey, W./Probst, W., Vegetationsanalytische Untersuchungen im Dünengebiet bei Babolsar (Kaspisches Meer, Iran), Bot. Jb. Syst. *94*, 1974 c, 96–113.

Frey, W./Probst, W., Gliederung der Vegetation und ihre Darstellung im Tübinger Atlas des Vorderen Orients (TAVO) und in den Beiheften zum Atlas, Beih. Tübinger Atlas Vorderer Orient, Reihe A (Naturwiss.) *1*, 1977, Wiesbaden.

Frey, W./Probst, W., Persische Golfküste. Mangrove. Karte A VI 10.7 des Tübinger Atlas des Vorderen Orients, 1986, Wiesbaden.

Gilli, A., Die Pflanzengesellschaften der Hochregion des Elbursgebirges in Nordiran, Beih. Bot. Centralbl. *59*, Abt. B, 1939, 317–344.

Ghorbanli, M./Lambinon, L., Premier aperçu de la zonation de la végétation halo-gypsophile du Lac Ghom, Lejeunea Rev. Bot. *92*, 1978, 4–23.

Hedge, I. C./Wendelbo, P., Patterns of distribution and endemism in Iran. Not. R. B. G. Edinb. *36*, 1978, 441–464.

Klein, J. C., Les groupements chionophiles de l'Alborz central (Iran). Comparaison avec leurs homologues d'Asie centrale, Phytocoenologia *10*, 1982, 463–486.

Kotschy, T., Die Vegetation des westlichen Elbrus in Nordpersien, Bot. Zeitschr. *4*, 1861, 1–15.

Kramer, W., Mittlerer Zagros (Iran). Vegetation. 1:600 000, Karte A VI 6 des Tübinger Atlas des Vorderen Orients, 1984, Wiesbaden.

Kürschner, H., Vegetation und Flora der Hochregionen der Aladağları und Erciyes Dağı, Türkei, Beih. Tübinger Atlas Vorderer Orient, Reihe A (Naturwiss.) *10*, 1982, Wiesbaden.

Kunkel, G., Beobachtungen zur Pflanzenwelt des Gueno-Massivs, Südiran, Philippea *2*, 1975, 317–321.

Kunkel, G., The vegetation of Hormoz, Qeshm and neighbouring islands (Persian Gulf area), Flora et Vegetatio Mundi *6*, 1977, 186 pp., Vaduz.

Leonard, J., Contribution à l'étude de la flore et de la végétation des deserts d'Iran, Fasc. 1–4, 1981–1984, Meise.

Melchior, H., Zur Pflanzengeographie des Elburs-Gebirges in Nord-Iran, Sitzb. Ges. naturforsch. Freunde Berlin, 1937, 55–71.

Mobayen, S., Aspect phytogéographique de l'Iran et particularités de végétation désertique du Lut, Env. Stud. (Teheran) *4*, 1975, 181–190.

Mobayen, S., Structure géobotanique du Loute, Acta Ecol. Iran *1*, 1976, 73–86.

Mobayen, S./Tregubov, V., Carte de la végétation naturelle de l'Iran, 1:2 500 000, Univ. Téhran, UNDP/FAO no. IRA 7, 1970.

Moore, P. D./Bhadresa, R., Population structure, biomass and pattern in a semidesert shrub, *Zygophyllum eurypterum* Boiss. & Buhse, in the Touran Protected Area in eastern Iran, Journ. Appl. Ecol. *5*, 1978, 719–727.

Mossadegh, A., Aperçu général sur les forêts de la Caspienne en Iran, Bull. Soc. forest. Franche-Compté prov. Est *4*, 1972.

Pontecorvo, G./Bokhari, M., Hedge-like habit of *Juniperus excelsa* at high altitude on the Southern Zagros Mountains in Iran, Proc. Roy. Soc. London, Ser. B *188*, 1975, 507–508.

PROBST, W., Vegetationsprofile des Elbursgebirges (Nordiran), Bot. Jb. Syst. *91*, 1972, 496–520.

PROBST, W., Beobachtungen zum Standortklima in verschiedenen Vegetationszonen des Elbursgebirges (Nordiran), Bot. Jb. Syst. *94*, 1974, 65–95.

PROBST, W., Zur Vegetationsgeschichte und Klimaentwicklung des Südkaspischen Waldgebietes (Nordiran). In: FREY, W./UERPMANN, H.-P. (Eds.), Beiträge zur Umweltgeschichte des Vorderen Orients, Beih. Tübinger Atlas Vorderer Orient, Reihe A (Naturwiss.) *8*, 1981, 26–39 Wiesbaden.

RASTIN, N., Vegetationskundliche Untersuchungen in Hochwaldresten der kaspischen Ebene, Phytocoenologia *11*, 1983, 245–289.

RECHINGER, K. H., Vegetationsbilder aus dem nördlichen Iran. In: KARSTEN, G./SCHENK, H. (Hrsg.), Vegetationsbilder *25* (5), 1939.

RECHINGER, K. H., Plants of the Touran Protected Area, Iran, Iran. Journ. Bot. *1*, 1977, 155–180.

RECHINGER, K. H./WENDELBO, P., Plants of the Kavir Protected Area, Iran, Iran. Journ. Bot. *1*, 1976, 23–56.

SCHWEIZER, G., Der Naturraum. In: GEHRKE, U./MEHNER, H. (Eds.), Iran, 1975, 13–39, Tübingen.

TABATABAI, M./JAVANSHIR, K., Iranian western oak forests. I. Kermanshaah- and Kordestan forests, Publ. Iran. Forest Serv. 1966, Teheran.

ZOHARY, M., On the geobotanical structure of Iran, Bull. Res. Counc. Isr. *11* D, 1963, 113 pp.

ZOHARY, M., Geobotanical foundations of the Middle East, 2 vols., 1973, Stuttgart, Amsterdam.

Acknowledgements

We wish to express our special thanks to the SFB 19 "Tübinger Atlas des Vorderen Orients" (DFG) and its spokesman Professor Dr. W. Röllig. We also thank our colleagues who worked with us in Iran, the late Professor Dr. P. Wendelbo and Professor Dr. M. H. Bokhari (now Šīrāz). For help in the field work we are grateful to our travel associates P. Braig, S. Bauer, Dr. R. Carle, Dr. W. Kramer, Dr. H. Kürschner and Dr. J. Schedler. We are grateful to Mr. M. Seidel (Tübingen) for translating the manuscript.

Prof. Dr. Wolfgang Frey, Institut für Systematische Botanik und Pflanzengeographie, Freie Universität Berlin, Altensteinstr. 6, D-1000 Berlin 33.

Prof. Dr. Wilfried Probst, Seminar für Biologie der Pädagogischen Hochschule, Mürwiker Str. 77, D-2390 Flensburg.

Photo 1 Cultivated land, remnants of the Hyrcanian lowland forest and Hyrcanian montane forest on the northern slopes of the Elburz Mountains.

Photo 2 Mangrove stands *(Avicennia marina)* on mud banks between Ǧazīre-ye Qešm and Bandar-e Ḫamīr.

Photo 3 *Cupressus sempervirens*-woodland near Ḥasanabād south of Čālūs.

Photo 4 *Juniperus excelsa*-woodland in the Kūh-e Kūrḫūd (between Gorgān and Boǧnūrd).

Photo 5 *Quercus brantii*-woodland between Yāsūǧ and Šisakt (Kūh-e Denā).

Photo 6 *Cold-deciduous scrub* in the mountains east of Čaman-e Bīd (province of Ḫorāsān).

Photo 7 *Open xeromorphic scrub* in the Kūh-e Garh southeast of Šīrāz.

Photo 8 "Bergmandel-Pistazien-Baumflur" in the sense of Bobek (1951) south of Sīrǧān (*Artemisia herba-alba* dwarf-scrub with *Pistacia* spec.-trees).

Photo 9 Semidesert shrubland and *Acacia aucheri*-“Pseudosavanna” on Ǧazīre-ye Qešm.

Photo 10 Semidesert shrubland on the coastal limestone hills near Bandar-e Ḫamīr. In the background Ǧazīre-ye Qešm.

Photo 11 *Open thorn-cushion formation* in a disturbed *Juniperus excelsa*-**woodland** near Čaman-e Bīd (province of Ḫorāsān). In the background the Kūh-e Kūrḫūd with *Juniperus excelsa*-woodland.

Photo 12 Semi-woody salt swamp (*Halocnemum strobilaceum* association) at the shore near Bandar-e Ḫamīr.

Photo 13 *High mountain scree formation* and *Alpine forb formation* on 'Alam Kūh (Central Elburz Mountains).

Photo 14 Devastated Hyrcanian montane forest east of Rūdbār, province of Gīlān. "Geschneitelte" *Quercus castaneifolia* trees.

Harald Kürschner

A physiognomical-ecological classification of the vegetation of southern Jordan

Contents

1. Introduction

Southern Jordan is one of the most interesting regions in the Middle East from the geobotanic point of view. Because this area is small, it is possible within a short time to traverse three regions which are completely different from one another geomorphologically and orographically:

- the Syro-African rift valley (Ghor depression)
- the western mountain ridge,
- the eastern sandy and gravelly desert plains and plateaus.

These regions show marked differences with respect to climate, geology, soil, flora and vegetation.

In addition to the "Flora Palaestina" (Zohary 1966, 1972, Feinbrun 1978) there are a number of recent floristic studies of the area (al-Eisawi 1982, 1983, a, b, Boulos 1977, Boulos et al. 1975, 1977, Boulos/al-Eisawi 1977 a, b, Boulos/Jallad 1975, Boulos/Lahham 1977 a, b, c). However, there is a lack of such investigations in the fields of vegetation ecology and plant sociology. Here one must reply upon the older (but in several cases excellent) surveys by Bardtke (1956), Eig (1938, 1946), Feinbrun/Zohary (1955), Kasapligil (1956), Long (1957), Zohary (1945), Zohary/

FEINBRUN (1951), ZOHARY/ORSHAN (1956/57) and ZOHARY/ORSHANSKY (1949). The only recent studies are a short article by ATKINSON/BEAUMONT (1971) on forests and a study of halophytes (Dead Sea area, al-Azraq) by FREY/KÜRSCHNER (1983).

The present investigation therefore endeavours to continue the study of vegetation in Jordan. It was written in the course of the work carried out by the Special Research Division 19, "Tübingen Atlas of the Middle East" (TAVO). With the aid of a vegetation map it gives a physiognomic-ecological description and classification of the vegetation in southern Jordan, along with a phytocoenological classification of natural regions. This makes it possible to compare the formations described here with other vegetation units in the Middle East and to relate the results obtained here with those of other studies (cf. previously published maps in the TAVO).

Here I would like to express my thanks to the Deutsche Forschungsgemeinschaft, which provided the funds for the necessary research trips through its support of SFB 19. I am especially grateful to my travelling companions, Prof. Dr. W. Frey (Berlin) and Dr. H. H. Hilger (Ulm).

2. Geological and climatic conditions

Owing to the aridity of the study area the character of the soils is strongly determined by geological conditions. The Geological Map of Jordan (BENDER et al. 1968) gives a good picture of these conditions. Archaic igneous rocks (Precambrian) up to 1500 m are particularly characteristic of the south-western part along Wādī l-ʿAraba. They consist of granites, gronodiorite, diorite, porphyrite, gneiss and mica schists. Like the massive, brownish Cambrian sandstones they yield red sands when subjected to weathering. Ordovician and Silurian fine sandstones and sandy shales are found mainly in south-eastern Jordan; they weather to white sands. The highlands of Edom and Moab are made up of limestones and marls of the Middle and Upper Cretaceous (Cenomanian, Santonian-Turonian) and of the Tertiary. The depressions and wadis in the east are filled with diluvial deposits, mainly gypsous chalky marls, loess sediments and eolian sands, and they empty into mudpans and large pans (Qāʿ al-Ğafr, Qāʿ al-Ḥafīra) with pelitic sediments. The eastern part is dominated by fluviatile gravels, mantle rocks and sandy limestones, which erode to form characteristic hammadas.

Jordan lies in the transition region between the zonobiome with moist winters and dry summers and the arid, temperate zonobiome with cold winters. However, it can still be seen as having a Mediterranean bioclimate (concentration of rains in the cold season, very dry summers). With the aid of the temperature-rainfall quotient Q (EMBERGER 1930, 1952) it can be subdivided into bioclimatic zones. These are helpful in reconstructing the potential natural vegetation and in determining the present-day extent of vegetation. Fig. 1 shows the calculated Q values for a number of climate stations, plotted against the mean minimum temperature of the coldest month. Three clusters can be observed:

- Group a: the stations of the western hill region with relatively high precipitation and temperate m,

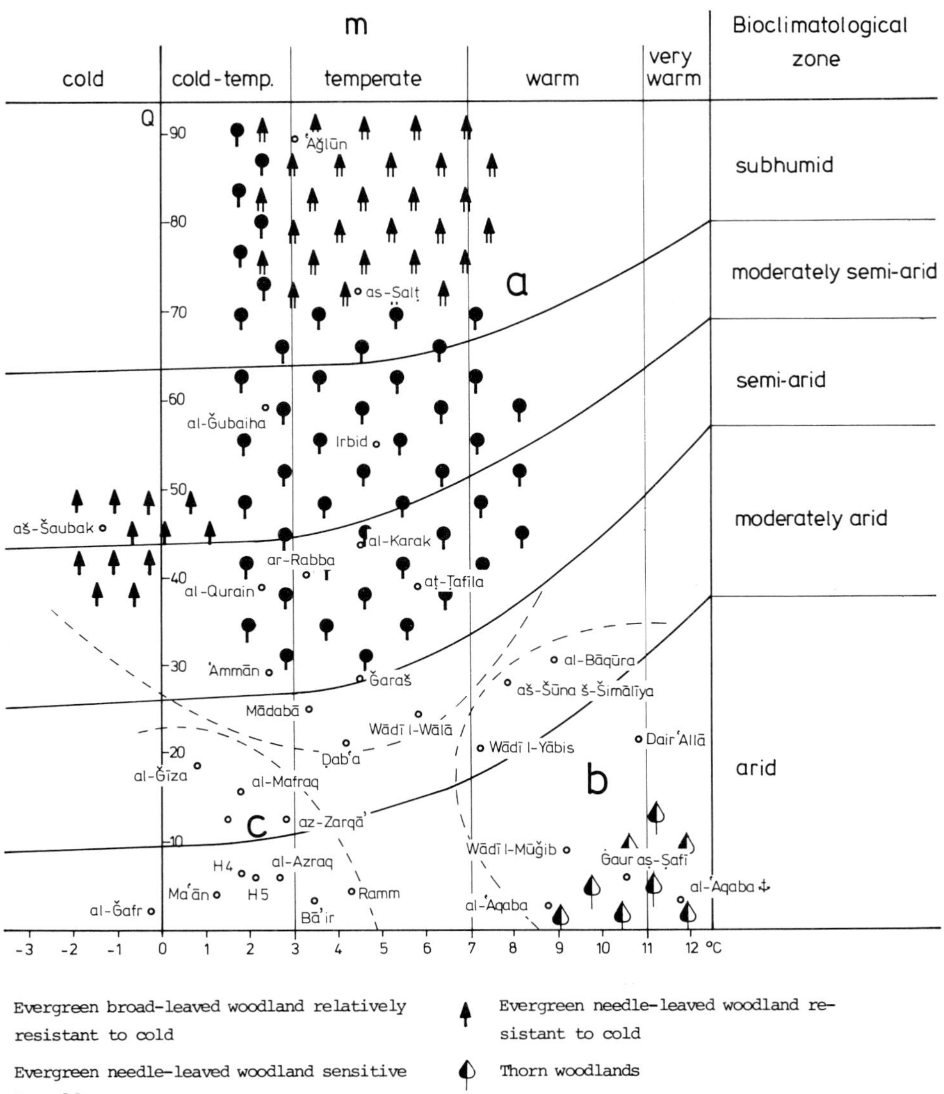

Fig. 1 Bioclimatic diagram of Jordan (quotient pluviothermique of EMBERGER) with corresponding positions of potential natural woodland formations.

- Group b: the stations of the rift valley with low precipitation and warm to very warm m,
- Group c: the stations of the eastern gravel and sand deserts with low precipitation and cold to cold-temperate m.

The woody-plant formations occurring in Jordan are also marked in the figure in accordance with their bioclimatic requirements.
Here it can clearly be seen which areas are potentially occupied by forests or woodland formations. In Jordan four bioclimatic zones (Map 1) can be distinguished, and these can be subdivided into different varieties depending on the value of m (Table 1).

I. Subhumid zone, temperate variety (V 1)

V 1	$3.0 < m < 4.3$	$72 < Q < 89$	$690 < P < 730$

II. Semi-arid zone, cold (V 1) and cold-temperate to temperate (V 2) variety

V 1	$m > -1.4$	$Q < 46$	$P < 300$
V 2	$2.4 < m < 5.8$	$28 < Q < 59$	$200 < P < 500$

III. Moderately arid zone, cold-temperate (V 1), temperate (V 2) and warm to very warm (V 3) variety

V 1	$0.7 < m < 2.8$	$12 < Q < 19$	$138 < P < 180$
V 2	$3.3 < m < 5.7$	$21 < Q < 25$	$216 < P < 270$
V 3	$7.3 < m < 10.9$	$21 < Q < 30$	$270 < P < 380$

IV. Arid zone, cold to cold-temperate (V 1), temperate (V 2) and warm to very warm (V 3) variety

V 1	$-0.3 < m < 2.6$	$2.5 < Q < 6.5$	$31 < P < 90$
V 2	$3.4 < m < 4.1$	$4.3 < Q < 5.0$	$50 < P < 60$
V 3	$8.8 < m < 12.0$	$2.7 < Q < 6.0$	$38 < P < 85$

Table 1 Bioclimatic zones of Jordan (m = mean of the minimum temperatures of the coldest month in °C, Q = quotient pluviothermique of Emberger, P = precipitation in mm/year).

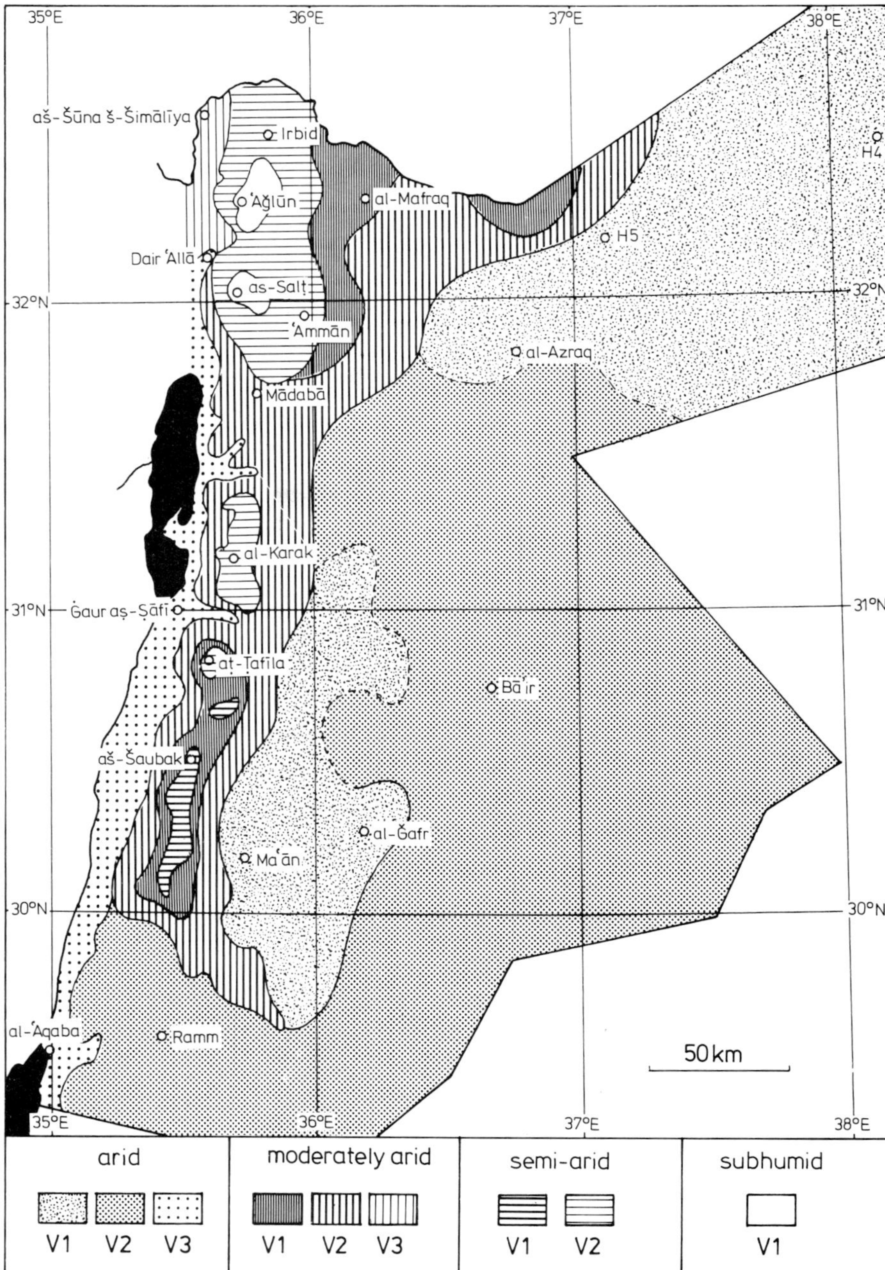

Map 1 Bioclimatic subdivision of Jordan

3. Methods

The vegatation was classified according to the physiognomic-ecological classification scheme used in the TAVO (FREY/PROBST 1977). The life forms of the characteristic taxa were taken as a starting point; these can be regarded as ecological equivalents of the various bioclimatic zones. By disregarding the species present and using a combination of physiognomic (mostly formal-morphological) and ecological criteria, a hierarchic system was constructed (theory of formations). Units relevant to the landscape were also included in this system.
The characteristic formations were recorded during two trips in 1980 (spring and summer). Information about the areas not visited was filled in by referring to published data (vegetation surveys) and by drawing inferences from the bioclimate, topographic indicators (e.g., altitude, morphology of the terrain) and geological and pedological data. The result is the vegetation map presented here.

4. Vegetation of southern Jordan (Map 3)

Jordan is of great interest in vegetation ecology because it is the meeting place of the Mediterranean, Irano-Turanian and Saharo-Arabian regions of the Holarctic kingdom and the Nubo-Sindian region of the Palaeotropical kingdom (Map 2, cf. also EIG 1931, 1938). For this reason there are conspicuous changes in the vegetation and in the composition of the flora over relatively short distances (within 30 km on the western slopes of the border mountains).
Thus it is natural that sharp boundaries occur between individual floristic regions only in exceptional cases. In general the areal system blend to form a continuum, so that objectified criteria must be employed for their demarcation. In Map 2 the criterion used was the change in the dominance of taxa from one floristic region to the next. The rosettes shown each represent a survey which was analysed. Rosettes similar in appearance were grouped together as floristic regions or as transition zones.
In the following the formations shown in Map 3 will be described according to the floristic regions to which they belong.

4.1 Vegetation units of the Mediterranean type

The formations of this type colonize mainly the western ridges and plateaus (Gilead, Ammon, Moab and northern Edom), where there is still a Mediterranean but also a continental climatic influence. Annual precipitation is between 400 and 700 mm, which is sufficient for the occurence of

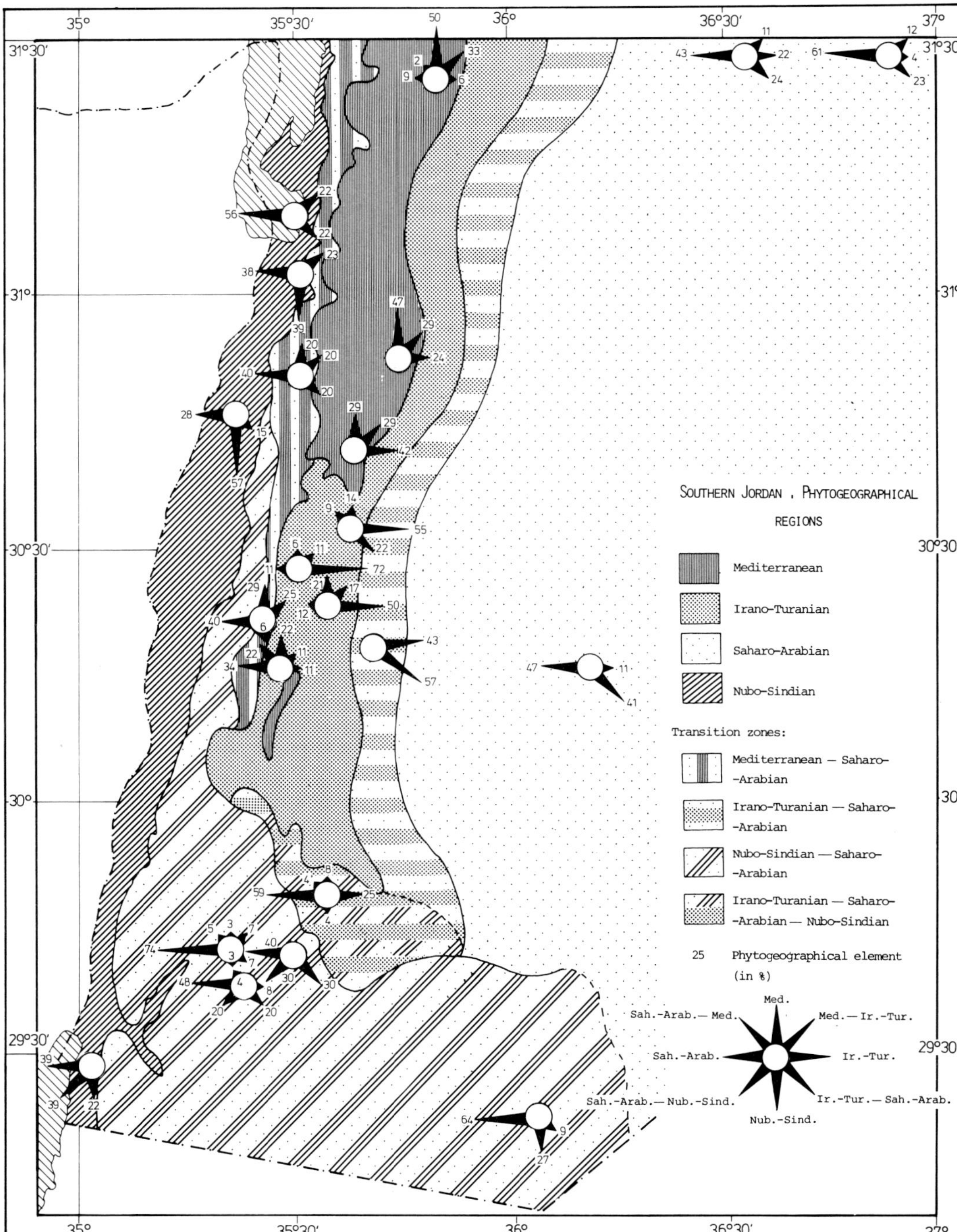

Map 2 Phytogeographical subdivision of southern Jordan (cf. EIG 1931, 1938, FEINBRUN/ZOHARY 1955).

woody-plant formations. Since precipitation drops off rapidly towards the east, these forest fragments are limited to a narrow zone. Today the main forest areas in the map section are in Fainān (south-west of aṭ-Ṭafīla) and between aṭ-Ṭafīla, aš-Šaubak and al-Ḥaišā. These forests grow along the western slopes of the old serir range (Ǧibāl aš-Šarāh) and form dense stands (cf. Musil, A. Arabia Petraea II, 1 Edom, 1907, p. 289 ff. and 299 ff.). Today they are highly degraded. At least four forest formations can be distinguished in recent times.

Evergreen broad-leaved woodland relatively resistant to cold (*3.22*, **Photo 1**)

Bioclimate: moderately arid to semiarid, cold-temperate to temperate

$0.77 < m < 5.8$ $12 < Q < 59$ $200 < P < 500$

Geology: The characteristic rocks are limestones and dolomites of the Cenomanian. Under the prevailing climatic conditions they have been transformed into fertile terra rossa soils.

Distribution: These evergreen woodlands, which can be classified as relatively resistant to cold, today form extensive shrub forests only in northern Jordan (Gilead, Ammon) at altitudes between 500 and 1000 m (Quercetum calliprini Feinbrun/Zohary 1955, *Quercus calliprinos–Pistacia palaestina* association Eig 1946). In northern Edom (1200–1600 m, Photo 1) they also occur in a xerophilous, very open variant. Here they form woodlands which have been affected by pastoral selection and which are characterized in the undergrowth by Irano-Turanian taxa adapted to aridity (*Quercus calliprinos–Juniperus phoenicea* association Feinbrun/Zohary 1955; this association appears to be artificial since the two characteristic species can be shown to have seperate habitats. *Quercus calliprinos* almost always colonizes limestone, while *Juniperus phoenicea* prefers sandstones and their weathering products. Cf. also Long 1957).

In historical times these oak forests in the northern Edom were much denser. The last major destruction phase was during the First World War, when large amounts of firewood were consumed between aš-Šaubak and ʿUnaiza by the Hiǧaz "forest railway" (Atkinson/Beaumont 1971), the vital supply line for the Turkish garnisons in Arabia (cf. Lawrence 1936). Today these forest remnants are regularly used as summer pasture by nomads from the eastern parts of the country.

Floristic composition:

Sample record 1: 15 km south of aš-Šaubak; alt. 1300 m; Santonian-Turonian limestones; exp. SW.; cov. 30%; date: 4 Apr. 1980.

Trees and shrubs: *Amygdalus korschinskii, Colutea istria, Quercus calliprinos.*

Chamaephytes: *Artemisia herba-alba, Euphorbia hierosolymitana.*

Hemicryptophytes: *Alyssum iranicum, Astragalus platyraphis, A. sanctus, Diplotaxis harra, Geranium tuberosum, Helianthemum vesicarium, Ranunculus damascenus, Scorzonera judaica.*

Therophytes: *Buglossoides tenuiflora, Carlina hispanica, Conringia orientalis, Holosteum glutinosum, Lasiopogon muscoides, Rochelia disperma, Veronica campylopoda, Viola occulta.*

Geophytes: *Gagea reticulata.*

Sample record 2: Near al-Ḥaišā, 10 km N. of Baṭrā; alt. 1400–1500 m; Cenomanian nodular limestones, locally dolomites; exp. W.; cov. 20–30%; date: 21 July 1980.

Trees and shrubs: *Cerasus microcarpa, Colutea istria, Ephedra campylopoda, Quercus calliprinos* var. *subglobosus.*

Chamaephytes: *Argyrolobium crotalarioides, Astracantha bethlehemitica, Centaurea damascena, Noaea mucronata, Pterocephalus pulverulentus.*

Hemicryptophytes: *Cousinia dayi, C. moabitica, Rubia tenuiflora, Stipa barbata.*

Therophytes: *Centaurea pallescens, Onopordum palaestinum, Picnomon acarna.*

Evergreen needle-leaved woodland resistant to cold (*3.32*, Photo 2)

Bioclimate: Semiarid, cold

$m < -1.4$ $Q < 46$ $P < 300$

Geology: Massive sandstones (undifferentiated, so-called Nubian sandstones) of the Cambrian formation are dominant.

Distribution: Needle-leaved woodland resistant to cold, consisting of *Juniperus phoenicea*, characterizes the western slopes of Edom between (600) 900 and 1500 m (Juniperetum phoeniceae Feinbrun/Zohary 1955). Here one finds an extremely drought-resistant variant of this circum-Mediterranean taxon, occurring on sandy weathering products of the Nubian sandstone (on thin soils, eroded sandy loams, Photo 2). In the shrub layer eastern Mediterranean taxa like *Colutea istria* are dominant, along with Saharo-Arabian descendants such as *Gomphocarpus sinaicus* and *Thymelaea hirsuta.* The herb layer proper is relatively poor in species. Large, more or less unbroken occurrences are found in the vicinity of Baṭrā, between Baṭrā and aš-Šaubak and near Ḍānā (Ǧabal al-ʿAṭāʾiṭa, Map 3).

Near the town of ar-Rašādīya (Ǧabal Sarāb) there is a variant, not shown on the map, in which *Cupressus sempervirens* is dominant. This is a relic that occurs on deep, sandy soils with a surface layer of humus (cf. Long 1957; Atkinson/Beaumont 1971).

Floristic composition:

Sample record: Near Baṭrā; alt. 800 m; massive brownish weathered Cambrian sandstones of continental environment; exp. W.; cov. 10–20%; date: 4 Apr. 1980.

Trees and shrubs: *Colutea istria, Crataegus aronia, Daphne linearifolia, Gomphocarpus sinaicus, Juniperus phoenicea, Pistacia palaestina, Thymelaea hirsuta.*

Chamaephytes: *Astragalus spinosus, Euphorbia hierosolymitana.*

Therophytes: *Galium hierochunticum, Parietaria alsiniifolia, Plantago afra, Reichardia tingitana.*

Mixed formation of *Evergreen needle-leaved woodland resistant to cold (3.32)* and *Xeromorphic dwarf-shrublands (9.4)*

Bioclimate: moderately arid, cold-temperate to temperate.

$0.7 < m < 5.7 \qquad 12 < Q < 25 \qquad 50 < P < 200$

Geology: cf. 3.32.

Distribution: In the western border mountains between Baṭrā and Ǧabal Mabrak. This mixed formation is characterized by very open, patchy and isolated *Juniperus* stands in sandstone cliffs. The undergrowth is dominated by *Artemisia herba-alba* and other Irano-Turanian dwarf-shrubs from the dwarf-shrub zone adjacent to the east (cf. FEINBRUN/ZOHARY 1955, Juniperetum with Irano-Turanian steppe associations). Similar transition formations have been reported from North Africa (Morocco, Tunisia, EMBERGER 1939, LONG 1957).

Floristic composition:

Cf. 3.32 and 9.4/21.21. Here however, the tree and shrub layers are opener and more patchy.

Debris and gravel desert (18.32 b)

Bioclimate: moderately arid, temperate.

$3.3 < m < 5.7 \qquad 21 < Q < 25 \qquad 200 < P < 500$

Geology: Mainly massive sandstones with red shales (Cambrian formation).

Distribution: The area below the *Juniperus phoenicea* woodlands in Edom (Baṭrā and surroundings), lying between 600 and 900 m, is characterized by a number of formations about which little is known with regard to extent and floristic composition.

What is most conspicuous are the fragments of a *Pistacia khinjuk-Ceratonia siliqua* community *(Cold-deciduous montane woodland with evergreen needle-leaved trees or shrubs)*, occurring only in gorges and wide crevices. This community is composed of taxa from the Mediterranean, Irano-Turanian and Saharo-Arabian regions. The rocks and screes are characterized by lithophilous and chasmophilous associations. In the wadis which dissect these ranges one of two communities is dominant, depending on the substrate. *Retama raetam* (Retametum roetami EIG 1946, Saharo-Arabian, psammophytic association) is restricted to sandy wadis, while *Nerium oleander* shrubs (Nerion oleandri FEINBRUN/ZOHARY 1955, Mediterranean hygrophytic association of tropical origin) grows in gravelly wadis with spring outlets. The many variations in this vegetation mosaic could not be shown on the map.

Floristic composition:

Sample record 1: Near al-Ḥaišā N. of Baṭrā; alt. 900 m; Massive Cambrian sandstones with gorges; exp. NE.; cov. low; date: 21 July 1980. (*Pistacia-Ceratonia* fragments in crevices and fissures).

Trees and shrubs: *Ceratonia siliqua, Crataegus aronia* var. *pontica, Daphne linearifolia, Ficus pseudosycomorus, Gomphocarpus sinaicus, Pistacia khinjuk* var. *glabra, Retama raetam, Rhamnus dispermus, Thymelaea hirsuta.*

Chamaephytes: *Achillea fragrantissima, Ononis natrix* ssp. *hispanica, Teucrium divaricatum.*

Hemicryptophytes: *Caralluma aaronis, Galium canum, Oryzopsis miliacea, Podonosma orientalis.*

Therophytes: *Delphinium peregrinum, Polygonum arenastrum, Stachys neurocalycina.*

Geophytes: *Urginea maritima.*

Sample record 2: Wādī Mūsā; alt. 700–900 m; gravel with spring horizons; exp. W.; cov. up to 90%; date: 23 July 1980 (*Nerium oleander* shrubs).

Trees and shrubs: *Celtis australis, Colutea istria, Gomphocarpus sinaicus, Nerium oleander, Pistacia khinjuk* var. *glabra, P. palaestina, Rubus sanguineus, Salix pseudo-safsaf.*

Hemicryptophytes: *Cyperus fuscus, Equisetum ramosissimum, Inula viscosa, Oryzopsis miliacea, Parietaria diffusa, Phragmites australis, Pulicaria inuloides, Tripteris vaillantii, Typha domingensis, Veronica anagallis-aquatica.*

Therophytes: *Conyza bonariensis, Digitaria sanguinalis, Eragrostis cilianensis, Lactuca serriola, Setaria verticillata, Tribulus terrestris.*

The major part of the Mediterranean-influenced region is taken up by cultivated land (rainfed cultivation) and batha communities (degraded maquis).

Mixed dwarf-scrub and herbaceous formations (9.3)

Bioclimate: moderately arid, semi-arid (so subhumid), temperate.

$2.4 < m < 5.8 \qquad 21 < Q < 59 \qquad 200 < P < 500$

Geology: Cenomanian limestones; clays, marls, limestones and dolomites of the Santonian-Turonian, which weathered to terra rossa and grey soils.

Distribution: In most of the Mediterranean-influenced region (Ammon, Moab, Map 3) batha communities are dominant (Poterietum spinosi mediterraneum EIG 1938, Salvietum graveolentis EIG 1946, syn. *Salvia domin-*

ica-Ballota undulata assoc. FEINBRUN/ZOHARY 1955). In the south and east they border on the Irano-Turanian dwarf-shrublands, and they are a xerix type of a Mediterranean maquis.
A common species in the deep wadi cuts of southern Moab (Wādī l-Mūǧib, Wādī l-Ḥaṣā) is *Sarcopoterium spinosum.* It mainly colonizes terra rossa soils of the Cenomanian and limestones of the Senonian (especially red arid soils of the semi-arid zone). On the ridges north of aṭ-Ṭafīla this Poterietum is mixed with *Artemisia herba-alba.* According to LONG (1957) this community is one of the most regressive stages of forest types of the semi-arid zone (association of *Quercus aegilops* var. *ithaburensis* and *Pistacia atlantica,* cf. 4.2). If grazing pressure diminishes, it becomes established all by itself.
Floristic composition:
Sample record 1: Wādī l-Mūǧib, N. of al-Karak; alt. 400 m; on Cenomanian limestones; exp. NW.; cov. up to 70%; date: 1 Apr. 1980 (Poterietum spinosi mediterraneum, spring aspect).
Chamaephytes: *Ajuga chia, Ballota undulata, Euphorbia hierosolymitana, Phagnalon rupestre, Salvia dominica, Sarcopoterium spinosum, Suaeda asphaltica.*
Hemicryptophytes: *Astragalus sanctus, Centaurea eryngioides, Crepis hierosolymitana, Diplotaxis harra, Eremostachys laciniatus, Malabaila secacul, Onosma orientalis, Paronychia argentea, Poa bulbosa, Ranunculus asiaticus, Scorzonera papposa, Umbilicus intermedius.*
Therophytes: *Ainsworthia trachycarpa, Asperula arvenis, Avena sterilis* ssp. *ludoviciana, Bromus lanceolatus* var. *lanuginosus, B. rubens, Carduus australis, Centaurea hyalolepis, Chaetosciadium trichospermum, Chrysanthemum coronarium, Crepis sancta, Cynosurus elegans, Echium judaeum, Galium hierochunticum, Hypecoum imberbe, Lamarckia aurea, Lamium moschatum, Lathyrus inconspicuus, Matthiola aspera, Minuartia picta, Nothobasis syriaca, Ononis sicula, Papaver subpiriforme, Parentucellia flaviflora, Plantago afra, Pterocephalus brevis, Reseda alba, Rumex cyprius, Salvia horminum, Sedum pallidum, Silene palaestina, Sinapis alba, Trifolium angustifolium, T. clusii, Trigonella schlumbergeri, Urtica pilulifera.*
Geophytes: *Allium erdelii, A. gasyunense, A. neapolitanum, Asphodeline lutea, Fritillaria persica, Gladiolus atroviolacea, Leopoldia tenuiflora.*
Sample record 2: Near al-Karak; alt. 600 m; Santonian-Turonian limestones; exp. W.; cov. 40%; date: 19 July 1980 (*Salvia dominica–Ballota undulata* assoc. fragments in *Non-irrigated arable land,* summer aspect).
Chamaephytes: *Ballota undulata, Noaea mucronata, Ononis natrix* ssp. *hispanica, Pterocephalus pulverulentus.*
Hemicryptophytes: *Alkanna strigosa, Anchusa strigosa, Carlina hispanica* ssp. *galilaea, Dianthus strictus, Echinops gaillardotii, Eryngium glomeratum, Galium humifusum, Gypsophila arabica, Heliotropium rotundifolium,*

Hypericum triquetrifolium, Ononis antiquorum var. *leiosperma, Scrophularia xanthoglossa.*
Therophytes: *Carthamus tenuis* ssp. *foliosus.*

Non-irrigated arable land (21.21)

Bioclimate: semi-arid, temperate.
$2.4 < m < 5.8$ $28 < Q < 59$ $200 < < P < 500$
Geology: Mainly Turonian limestones.
Distribution: The upland plains of Moab and Edom (cf. also 9.4/21.21) have today been largely turned into cultivated land (dry farming of grain). Here there are a number of typical segetal plant communities. According to Feinbrun/Zohary (1955) the segetal associations of the unplanted fields can be distinguished geographically:
- *Ononis leiosperma–Phlomis orientalis* association: Ammon, between ʿAmmān and Mādabā.
- *Ononis leiosperma–Phlomis pungens* association: Moab, between Mādabā and al-Karak and parts of northern Edom.

Mixed formation of *Non-irrigated arable land (21.21)* and *Mixed dwarf-scrub and herbaceous formations (9.3)*

Distribution: The distribution of this mixed formation cannot be shown on the map in detail. Throughout the highlands of Moab there is an alternation of batha fragments and abandoned cultivate land.

4.2 Vegetation units of the Irano-Turanian type

The Irano-Turanian region characterizes large areas of Edom and extends in a narrow band to Moab in the north (Map 2). This area clearly differs from the one described above owing to the drop in annual precipitation down to 150–400 mm and to the frequent frosts in the winter months. Much of the Irano-Turanian region in Jordan is today under cultivation, with the result that there are only a few remnants of the potential natural vegetation (woodlands, "steppe forests" and dwarf-shrub formations).

***Cold-deciduous broad-leaved woodlands without evergreens** (4.2,* **Photo 3)**

Bioclimate: moderately arid, cold-temperate to temperate.
$0.7 < m < 5.7$ $21 < Q < 25$ $200 < P < 270$
Geology: Limestones and dolomites, chalky marls and clays of the Cenomanian and Santonian-Turonian.
Distribution: This formation was formed by steppe forests, which were formerly more extensive, in the transition zone to the dwarf-shrub formations. Today it occurs as a completely devastated remnant in the form of a narrow strip along the western hill ranges (from aṭ-Ṭafīla, Ǧabal al-

ʿAṭaʾiṭa, aš-Šaubak to Ǧabal Mabrak, Map 3), where it grows on loess and brown and grey Sierozems with a medium lime content (weathering products of the Cenomanian limestones). Generally the only remaining arboreal representative is *Pistacia atlantica* (Photo 3). According to Long (1957) these *Pistacia* tree meadows ("steppe forests", Pistacietum atlanticae Feinbrun/Zohary 1955) are an impoverished association of *Quercus aegilops* var. *ithaburensis* and *Pistacia atlantica*, and not Pistacietum atlanticae as described by Feinbrun/Zohary (1955).

In the area adjoining the steppe *Quercus aegilops* var. *ithaburensis* is becoming increasingly rare; today it has disappeared from the entire Moab plateau. Long, judging from edaphic conditions, infers that "the entire plateau of Mādabā, where about 95% of the available land is now cultivated, was covered by a mixed forest of *Quercus aegilops* var. *ithaburensis* and *Pistacia atlantica* some thousands of years ago."

Floristic composition:

Sample record: Near Ǧabal al-ʿAṭāʾiṭa; alt. 1400 m; Cenomanian limestones; exp. SW.; cov. 5%; date: 20 July 1980.

Trees and shrubs: *Amygdalus korschinskii, Crataegus aronia* var. *pontica, Pistacia atlantica.*

Chamaephytes: *Artemisia herba-alba, Majorana syriaca.*

Hemicryptophytes: *Cousinia moabitica, Phlomis kurdica.*

Therophytes: *Centaurea iberica.*

Mixed formation of *Xeromorphic dwarf-shrublands (9.4)* and *Non-irrigated arable land (21.21*, Photo 4)

Bioclimate: moderately arid, cold-temperate to temperate.

$0.7 < m < 5.7$ $21 < Q < 25$ $200 < P < 270$

Geology: Limestones, locally sandy, marls and dolomites mainly of the Santonian-Turonian. Greyish steppe soils (Sierozems) are dominant.

Distribution: Dwarf-shrublands ("steppes", Artemisietum herbae-albae Eig 1938, 1946, Photo 4) cover large areas in Edom and Moab between 1200 and 1500 m, and they are an excellent indicator of rainfed cultivation. The areas which they originally occupied are therefore now for the most part cultivated (dry farming and the use of a two-year rotation; cultivated fallow followed by wheat). In areas which have been degraded by grazing *Noaea mucronata* (Noëetum mucronatae Eig 1946) takes over, whereas on low-lying loess soils and in depressions the Artemisietum is replaced by an *Anabasis syriaca (A. haussknechtii)–Poa sinaica* association (Eig 1946, Feinbrun/Zohary 1955) – South of Ǧabal Mabrak, between Maʿān and Ra's an-Naqb, a subassociation with *Haloxylon articulatum* (Haloxylonion articulati Eig 1946) is dominant. This subassociation merges to the east with the more Chenopodiaceae-rich dwarf-shrublands (cf. 9.6 a) in the area of contact with the reg formations.

Floristic composition:
Sample record: Near aš-Šaubak; alt. 850 m; Santonian limestones with Sierozems; exp. NE.; cov. 60%; date: 10 Apr. 1980.
Chamaephytes: *Artemisia herba-alba, Astragalus spinosus, Noaea mucronata, Ononis natrix* ssp. *natrix.*
Hemicryptophytes: *Alkanna strigosa, Alyssum iranicum, Astragalus platyraphis, Carex pachystylis, Poa sinaica, Ranunculus damascenus, Scorzonera judaica, Tragopogon collinus.*
Therophytes: *Alyssum marginatum, Androsace maxima, Astragalus cruciatus, A. tribuloides* var. *minutus, Bassia eriophora, Ceratocephalus falcatus, Crepis sancta, Descurainia sophia, Eremopyrum distans, Lactuca undulata, Lasiopogon muscoides, Lophochloa phleoides, Malcolmia africana, M. crenulata, Matthiola livida, Minuartia picta, Nardurus subulatus, Reseda lutea, Roemeria hybrida, Rumex dentatus, Torularia torulosa, Valerianella dufresnia, Veronica campylopoda, Ziziphora tenuior.*
Geophytes: *Gagea reticulata, Tulipa polychroma.*

Xeromorphic, very open dwarf-shrublands (9.6)

Bioclimate: moderately arid, cold-temperate.
$0.7 < m < 3.0$ $12 < Q < 25$ $130 < P < 270$
Geology: Limestones and dolomites of the Santonian–Turonian, locally sandy marl, intermixed with fluviatile gravel and Quaternary deposits. Mainly greyish and brownish Sierozems are dominant.
Distribution: Because of low precipitation the dwarf-shrublands thin out towards the east, becoming very open formations in which Chenopodiaceae determine the physiognomy (association of *Artemisia herba-alba* and *Anabasis aphylla* Long 1957, 9.6 a, Map 3). In the area of contact with the stone and gravel deserts *Artemisia herba-alba* diminishes more and more, giving way to a *Hammada scoparia–Salsola vermiculata* association (Feinbrun/Zohary 1955, association of *Haloxylon articulatum–Salsola villosa* Eig 1946, 9.6 b, Map 3), which in turn merges into the Saharo-Arabian formations.
Floristic composition:
Comp. 9.4/21.21; in addition *Aellenia lancifolia, Anabasis setifera, A. syriaca, Halotis pilosa, Hammada scoparia, Salsola inermis* and *Traganum nudatum* co-dominant.

4.3 Vegetation units of the Saharo-Arabian type

The communities of the Saharo-Arabian region colonize four main biotopes, which are ecologically very distinct:
– the hammadas in eastern Jordan, which receive less than 100 mm of precipitation per year,

- the sand deserts in the south and south-east (here heavily interspersed with species of the Nubo-Sindian element),
- the depressions, wadis and salt pans in the east,
- the lower hill zone in Wādī l-ʿAraba.

Precipitation in this region is less than 150 mm per year and is generally erratic. In al-ʿAqaba, for example, 65.7 mm of precipitation fell within 24 hours in February, 1975, whereas during the entire period from September 1974 to September 1975 (with the exception of February) only 10 mm was recorded (Boulos/Lahham 1977 a). The mean annual precipitation is 42 mm. This irregularity in the supply of moisture creates very difficult conditions for plant growth. On the other hand, it repeatedly leads to the formation of rich episodical desert forb communities.

Halophytic semi-woody shrub communities (14.1)

Bioclimate: arid, warm to very warm.

$8.8 < m < 12.0$ $2.7 < Q < 6$ $38 < P < 85$

Geology: Marls, clays and gypsum with clastic intercalations ("Lisan Marls").

Distribution: On hydro- and subhydrohalopedic soils (halobiomes with Solonchaks, gypsum-rich Lisan marl) in Wādī l-ʿAraba and in the depressions near al-Ǧafr (Qāʿ al-Ǧafr). These Saharo-Arabian-influenced halophyte communities are divided on the basis of edaphic conditions into a hydrohalophytic and a xerohalophytic series (Atriplicetum halimi, Nitrarietum retusae, Phragmitetum communis salinum, Salsoletum tetrandrae, Suaedetum palaestinae Eig 1946, Zohary/Orshansky 1949, Feinbrun/Zohary 1955, Zohary/Orshan 1956/57).

Because there is generally a pronounced edaphic gradient, they have a characteristic zonation. Examples of this from Wādī l-ʿAraba have been described by Zohary (1945) and Frey/Kürschner (1983). The succession on the Dead Sea (al-Baḥr al-Mayyit) north of Ġaur aṣ-Ṣāfī is as follows: the innermost, polyhaline region, which is inundated for a long time in the spring, is colonized by *Salicornia europaea.* Associates are *Arthrocnemum macrostachyum* and *Cressa cretica.* This first zone is followed by a *Sudaea aegyptiaca* zone, which is penetrated by *Tamarix tetragyna* shrubs in moist, occasionally inundated erosion channels. In the dryer oligohaline zones beyond the flooded area *Seidlitzia rosmarinus* is dominant (hygric and salt-dependent zonation). *Juncus rigidus* (Juncetum arabici) and *Phragmites australis* characterize mesohaline gullies and the outlets of brackish springs. These species form the transition to the hydrophytic communities.

Floristic composition:

Sample record: 15 km S. of al-Mazār (al Baḥr al-Mayyit); alt. –390 m; "Lisan Marls"; cov. 30–50%; date: 24 July 1980.

Shrubs: *Arthrocnemum macrostachyum, Nitraria retusa, Seidlitzia rosmarinus, Tamarix tetragyna.*

Hemicryptophytes: *Aeluropus litoralis, Cressa cretica, Juncus rigidus, Phragmites australis, Salsola jordanicola.*

Therophytes: *Salicornia europaea, Suaeda aegyptiaca.*

Mixed formation of *Deserts (18.)* and *Halophytic semi-woody shrub communities (14.1)*

Bioclimate: arid, temperate.

$3.4 < m < 4.1 \qquad 4.3 < Q < 5 \qquad 40 < P < 60$

Geology: Pelitic sediments in mudpans interbedded with fluviatile gravel and lacustrina limestones and clay, deposited after the Pliocene regression.
Distribution: In the wadis, gullies, pans and depressions in the south-eastern and eastern parts of the study area. Owing to the great edaphic variability these habitats can be classified only with difficulty. The mudpans and larger wadis are generally plantless in the middle, whereas the areas on the margins are in some cases dominated by halophytes, drift-line communities or annual meadows (after spring rains).
Floristic compositions:

Sample record 1: Qāʿ al-Ǧafr, SW. of al-Ǧafr; alt. 720 m; sandy, gravelly wadi; cov. 5%; date: 11 Apr. 1980.

Shrubs: *Atriplex halimus, Haloxylon persicum, Retama raetam, Salsola tetrandra, S. vermiculata, Seidlitzia rosmarinus.*

Chamaephytes: *Anabasis articulata, Halogeton alopecuroides, Helianthemum sessilifolium, Teucrium polium.*

Hemicryptophytes: *Cynodon dactylon, Fagonia bruguieri, Launaea nudicaulis, Peganum harmala, Scorzonera pusilla.*

Therophytes: *Astragalus tribuloides* var. *minutus, Erodium desertii, Hordeum marinum* var. *pubescens, Malva parviflora, Plantago ciliata, Schismus barbatus.*

Sample record 2: Wadī l-Ġadaf, S. of al-Azraq; alt. 520 m; fluviatile gravel and sand; cov. 20–30%; date: 7 Apr. 1980 (rich in annuals after rainfall in spring).

Chamaephytes: *Artemisia herba-alba, Astragalus spinosus, Salsola longifolia.*

Hemicryptophytes: *Anchusa italica, Astragalus acinaciferus, Onobrychis ptolemaica, Poa sinaica, Salvia lanigera.*

Therophytes: *Astragalus corrugatus, A. tribuloides* var. *leiocarpus, Calendula arvensis, Centaurea ammocyanus, Erodium desertii, Erucaria boveana, Filago pyramidata, Gastrocotyle hispanica, Helianthemum ledifolium, Lolium rigidum, Matthiola longipetala, Phalaris minor, Plantago amplexicaule* ssp. *bauphula, P. coronopus* ssp. *commutata, Reseda decursiva, Rumex vesicarius, Savignya parviflora, Sisymbrium septulatum, Trigonella moabitica, Tripleurospermum auriculatum, Vicia monacantha.*

Geophytes: *Bellevalia stepposum* var. *edumaea, Iris sisyrinchium.*

***Sand desert with shrubs** (**18.21**, **Photo 5, 6**)*

This Saharo-Arabian-influenced formation occurs in three varieties, which are geographically distinct.

Variety 1 (18.21 a, Map 3)
Bioclimate: arid, temperate.
$3.4 < m < 4.1$ $4.3 < Q < 5$ $50 < P < 60$
Geology: Eolian sands, fine sandstones and sandy shales of marine environment (Silurian-Ordovician).
Distribution: The sand fields and sand dunes (psammobiomes) in the south-east (region around al-Mudawwara) are characterized by psammophytic associations in which *Haloxylon persicum* plays a dominant role ("Saxaul forests", "Ghada trees"). This chenopod is important as a dune stabilizer and as a source of firewood for the inhabitants of the desert. Sociologically these communities can be grouped with the Haloxylotea salicornici EIG (Haloxylion persici EIG 1946, association of *Haloxylon salicornicum* and *Artemisia judaica* LONG 1957).
Floristic compositions:
Sample record: 30 km N. of al-Mudawwara; alt. 720 m; eolian sands; exp. S.; cov. less than 5%; date: 14 Apr. 1980.
Shrubs: *Haloxylon persicum.*
Chamaephytes: *Artemisia judaica, Asteriscus graveolens.*
Hemicryptophytes: *Stipagrostis plumosa.*
Therophytes: *Astragalus schimperi, Centaurea pallescens, Cleome amblyocarpa, Plantago ciliata, Savignya parviflora, Zygophyllum simplex.*

Variety 2 (18.21 b, Map 3)
Bioclimate: cf. variety 1.
Geology: Eolian sands and weathering products of Cambrian and Ordovician sandstones.
Distribution: A variant of the Haloxylotea occurs in Wādī Ramm and surroundings. The dominant species is *Hammada salicornica* (Haloxylion salicornici EIG 1946, Hammadetea salicornica ZOHARY 1973, Photo 5), which prefers deep, flat beds of fine sand. This variant is also characterized by the occurrence of a species-rich therophytic vegetation. According to studies in the Sahara by DUBIEFF (1959) the therophytes in these arid regions with episodical precipitation require rainfall of at least 5 mm/day (with possible intervalls of 5–10 years). This abundance of annuals could not be observed in the area of al-Mudawwara in the spring of 1980.
Floristic composition:
Sample record: Wādī Ramm; alt. 720 m; sandy plains; exp. S.; cov. up to 40%; date: 13 Apr. 1980 (Hammadetea salicornica rich in Therophytes).
Shrubs: *Haloxylon persicum.*

Chamaephytes: *Anabasis articulata, Gymnocarpus decandrum, Hammada salicornia.*
Hemicryptophytes: *Ajuga chia, Diplotaxis harra, Fagonia glutinosa, Kickxia floribunda, Morettia philaeana, Onobrychis ptolemaica, Stipagrostis obtusa.*
Therophytes: *Aegilops cylindrica, Anisosciadium isosciadium* var. *idumaeum, Astragalus gyzensis, Atriplex dimorphostegia, Bassia muricata, Crucianella membranacea, Cutandia memphitica, Eremobium aegyptiacum, Erodium laciniatum* var. *pulverulentum, Gastrocotyle hispida, Hyoscyamus pusillus, Ifloga spicata, Linaria haelava, Lotus halophilus, L. peregrinus, Matthiola aspera, Medicago truncatula, Ononis serrata, Picris cyanocarpa, Pimpinella cretica, Plantago cylindrica, Scabiosa eremophila, Schimpera arabica, Schismus barbatus, Silene arabica, Stigmatella longistyla.*
Geophytes: *Allium sindjarense, Asphodelus refractus, Dipcadi erythraeum, Pancratium sickenbergeri.*

Variety 3 (18.21 c, Map 3)
Bioclimate: arid, warm to very warm.
$8.8 < m < 12.0$ $2.7 < Q < 6$ $38 < P < 85$
Geology: Eolian sands, Quaternary river and wadi deposits, weathering products of Precambrian granites and Cambrian sandstones.
Distribution: Wādī l-ʿAraba. Sociologically this variety consists of relatively species-poor psammophytic communities of the Haloxylo-Retametalia retami Zohary 1973 (Haloxylonetum persici Eig 1946), which are confined to the dunes and sand fields of Wādī l-ʿAraba (Photo 6). Associates are *Calligonum comosum* and *Retama raetam.*
Floristic composition:
Sample record: S. of Ġaur aṣ-Ṣāfī; alt. –200 m; sandy plain; exp. W.; cov. less than 5% date: 24 July 1980.
Shrubs: *Calligonum comosum, Haloxylon persicum, Retama raetam.*
Chamaephytes: *Hammada salicornica, Moltkiopsis ciliata.*
Hemicryptophytes: *Polycarpaea repens, Stipagrostis raddiana.*
Therophytes: *Bassia muricata, Zygophyllum simplex.*

Rock desert (*18.31*, Photo 7)

Rock and gravel deserts (hammadas, regs) characterize the landscape of eastern Jordan, but they also occur in some places in Wādī l-ʿAraba. These hammadas (a comprehensive term for rock deserts with desert varnish) can develop only under extreme desert conditions.
There is usually silex or debris of calcareous crusts, and the soil surface is covered at least 75% by stones. According to Evenary/Orshansky (1948), four developments can be distinguished:

- Grit hammada: Poor in water-soluble salts. Vegetation is usually confined to depressions and gullies (mode contracté Monod 1954, restricted type).

- Sand hammada: Boulder pavements with secondary sand drifts. Very poor in water-soluble salts. The sand over the compact crust enables the upper layers to be quickly saturated after rain showers, and the water can be taken up by therophytes with shallow roots. This surface water disappears quickly through absorption by plants and evaporation. Thus there is not enough available for hemicryptophytes with their luxuriant shoot systems (FRANKENBERG 1978).
- Non-saline hammada: Low salt content, large fraction of marls and clays with poor water storage (extremely dry locations).
- Salt hammada: Rich in gypsum and salts. Plant growth often possible only after leaching of the salts from the upper soil layers (annual meadows after spring rains).

Variety 1 (18.31 a, Map 3)

Bioclimate: arid, cold-temperate to temperate, warm.
$-0.3 < m < 4.1$ $2.5 < Q < 5$ $30 < P < 60$

Geology: Sandy and conglomeratic limestones (Lower Tertiary and Upper Cretaceous), locally marls and Quaternary mantle rocks.

Distribution: The area east of the road from al-Qaṭrāna to Maʿān is dominated by nearly salt-free hammadas which are sterile on account of aridity and mechanical influences (wind erosion).

Another reason for the sterility of these boulder pavements (lithobiomes) is their low water-holding capacity. Characteristically they have a sparse vegetation consisting mostly of Chenopodiaceae; after rain showers, however, the physiognomy can be determined for a short time by annuals. Sociologically these formations belong to the Anabasidetea articulati FEINBRUN/ZOHARY 1955 (*Anabasis articulata* assoc. LONG 1957), which has been described as comprising a number of alliances, depending on edaphic and topographic conditions (Anabasidion articulati, Chenoleion arabicae, Gymnocarpion fruticosi, Salsolion villosae EIG 1946, ZOHARY/ORSHAN 1956/57). In some places this Anabasidetea is so degraded that vegetation is almost non-existent and the character species *Anabasis articulata* is absent over wide areas (e.g. between Maʿān and al-Ǧafr, Photo 7). According to LONG (1957) parts of these hammadas were originally covered by sub-shrubs (and in some places even by stands of *Acacia tortilis*, a recent example being Wādī l-Muṣawwal, subassociation Anabasidetosum articulatae Acacietosum ZOHARY 1945). From archaeo-zoological data and discoveries of animal bones LONG concludes that these areas had a rich "desert fauna" (especially gazelles). However, this formation, which is much richer in woody plants, is correlated with higher winter temperatures (less continentality) and is now probably no longer present.

Floristic composition:

Sample record: W. of al-Azraq; alt. 520 m; Quaternary mantle rocks, reg formation; cov. up to 10%; date: 3 Apr. 1980 (formation rich in annuals).

Hemicryptophytes: *Astragalus tarchoniticus, Erodium hirtum, Poa sinaica.*

Therophytes: *Aaronsohnia factorovskyi, Adonis aestivalis, Aizoon hispanicum, Arnebia linearifolia, Asteriscus pygmaeus, Astragalus palaestinus, A. tribuloides* var. *minutus, Centaurea sinaica, Erodium bryoniifolium, E. desertii, Erucaria boveana, Filago desertorum, Gastrocotyle hispida, Gymnarrhenia micrantha, Helianthemum ledifolium, Herniaria hirsuta, Lappula spinocarpos, Limonium thounii, Matthiola longipetala, Picris damascena, Plantago coronopus* ssp. *commutatus, P. ovata, Pteranthus dichotomus, Reichardia tingitana, Roemeria hybrida, Schismus arabicus, Spergularia diandra, Stipa capensis, Tetradiclis tenella, Torularia torulosa, Trigonella stellata.*

Geophytes: *Allium rothii, Bellevalia zoharyi.*

Variety 2 (18.31 b, Map 3)

Bioclimate: cf. variety 1

Geology: cf. cariety 1, dense boulder pavements with hardened marly-clayey interspaces.

Distribution: E. of al-Ǧafr, region around Bā'ir. Vegetation confined to gullies. Type: non-saline hammada.

Floristic composition:

Sample record: S. of al-Azraq, area of Wādī l-Ġadaf and Wādī l-Maḫrūq; alt. 520 m; Non-saline hammada; cov. up to 10%; date: 7 Apr. 1980.

Shrubs and chamaephytes: *Anabasis articulata, Asteriscus graveolens, Atriplex halimus, Chenolea arabica, Ephedra alte, Helianthemum stipulatum, Salsola tetrandra, Seidlitzia rosmarinus, Zilla spinosa.*

Hemicryptophytes: *Astragalus palaestinus, Ferula communis, Launaea nudicaulis, Onobrychis ptolemaica, Stipa parviflora.*

Therophytes: *Althaea ludwigii, Asteriscus pygmaeus, Astragalus tribuloides* var. *minutus, Bromus commutatus, Centaurea sinaica, Eremopyrum bonaepartis, Erodium bryoniifolium, Gymnarrhenia micrantha, Herniaria hirsuta, Lappula spinocarpos, Matthiola livida, Picris damascena, Plantago ovata, Reboudia pinnata, Spergularia diandra, Tetradiclis tenella, Torularia torulosa.*

Geophytes: *Bellevalia mocheovii, Iris sisyrinchium.*

Variety 3 (18.31 c, Map 3)

Bioclimate: cf. variety 1.

Geology: cf. variety 1, boulder pavement with large sand drifts in some places.

Distribution: Shown on the map only north-east of al-Qaṭrāna but certainly present elsewhere in the study area. Type: sand hammada. No information is available about the floristic composition, but it is probable that annuals of varieties 1 and 2 here too determine the physiognomy after periods of heavy precipitation.

Variety 4 (18.31 d, Map 3)
Bioclimate: arid, warm to very warm.
$8.8 < m < 12.0$ $2.7 < Q < 6$ $38 < P < 80$
Geology: Fluviatile gravel (Pleistocene) and mantle rocks.
Distribution: This variety is limited to Wādī l-ʿAraba. It is usually found on boulder pavement with onblown sand (type: sand hammada), where *Hammada scorparia* is co-dominant. Owing to the frequent occurrence of arboreal components (*Acacia raddiana, A. tortilis, Lycium shawii* and others) this variety can be put in the subassociation Anabasidetum articulatae Acacietosum sensu ZOHARY (1945), which is characteristic of Wādī l-ʿAraba.

Debris and gravel desert (*18.32 a*, Photo 8)

Bioclimate: arid, cold-temperate to temperate.
$-0.3 < m < 4.1$ $2.5 < Q < 5$ $30 < P < 60$
Geology: Precambrian granites, granodiorites, quartz diorites and massive, locally bedded Cambrian sandstones. Lithobiome with predominantly physical weathering (cracking from heat, wind erosion).
Distribution: The relict mountains and inselbergs (Photo 8) that rise from the sand plains in the south (Ǧabal Ramm, Ǧabal Umm Sahm etc.) and the mountain ranges north of al-Aqaba (Ǧabal Aḥmad al-Bāqir) must be characterized as *Debris and gravel desert.* Woody plants are largely absent here. At spring horizons, which occur where permeable Cambrian sandstone meets Precambrian granites with a marl layer, one may occasionally find *Ficus pseudosycomorus, Gomphocarpus sinaicus* and *Rhamnus dispermus.* Within this formation lithophytic and chasmophytic associations, rich in chamaephytes and hemicryptophytes are dominant (Halogetonetum alopecuroidis, Varthemietum iphionoidis, association of *Cheilanthes fragrans – Ceterach officinarum* and others EIG 1946). Few details are yet known about them.
Floristic composition:
Sample record: Ǧabal Ramm, foot of mountain and slope; alt. 720–1000 m; massive, brownish weathered sandstones; scree formation: exp. NE.; cov. 10%; date: 14 Apr. 1980.
Shrubs: *Capparis spinosa* var. *aravenis.*
Chamaephytes: *Aellenia lancifolia, Anabasis articulata, Aerva persica, Asteriscus graveolens, Ephedra alte, Farsetia aegyptiaca, Halogeton alopecuroides, Helianthemum kahiricum, H. sancti-antonii, Lavandula coronopifolia, Reaumuria hirtella, Stachys aegyptiaca, Traganum nudatum.*
Hemicryptophytes: *Hyparrhenia hirta, Kickxia acerbiana, Reseda muricata.*
Therophytes: *Bromus rubens, Callipeltis aperta, Caylusea hexagyna, Hippocrepis bicontorta, Koelpinia linearis, Medicago truncatula, Papaver decais-*

nei, Paracaryum intermedium, Parietaria alsinifolia, Plantago maris-mortui, Scabiosa porphyroneura, Telephium sphaerospermum, Tripleurospermum auriculatum.

Mixed formation of *Debris and gravel desert (18.32)* and *Xeromorphic dwarf-shrublands (9.4)*

Bioclimate: moderately arid, temperate to warm.

$3.3 < m < 10.9$ $21 < Q < 30$ $200 < P < 300$

Geology: Massive sandstones with red shales (Cambrian), Quartz-porphyry, clastics in volcanic matrix, and granites of the Precambrian, intermixed with sandy limestones of the Upper Cretaceous.

Distribution: The lower hill zone (up to 600–700 m) along Wādī l-ʿAraba is difficult to classify, and large parts of it are unknown. In this region dwarf-shrublands dominate which vary considerably in composition and floristic origin. No information is available about woody-plant formations in wadis and on pediments. The sucession of vegetation could be observed only at the few locations which are easily accessible. On the slopes in Wādī l-Karak, for example, the Mediterranean batha communities (cf. 9.3) are followed first by a belt of *Salsola-Reaumuria hirtella* (Reaumurietum palaestinae Eig 1946) and then by a *Suaeda asphaltica* zone (Suaedetum asphalticae Eig 1946) as one descends further into Wādī l-ʿAraba. The wadis are characterized by *Retama raetam.* In the larger wadis the first *Acacias* occur already at 600 m. The wadi outlets in Wādī l-ʿAraba are occupied by dense stands of *Phoenix dactylifera, Desmostachya bipinnata* and *Saccharum* spp. (Photo 9).

Mixed formation of *Debris and gravel desert (18.32)* and *Sand-desert with dwarf-shrubs (18.22)*

Bioclimate: arid, temperate.

$3.4 < m < 4.1$ $4.3 < Q < 5$ $50 < P < 60$

Geology: Massive, whitish weathered Ordovician sandstones.

Distribution: North and north-east of al-Quwaira (between Ra's an-Naqb and al-Quwaira) the landscape between 800 and 1200 m is characterized by white sands and wind-eroded sandstone peaks. The sandy slope at the foot of the sandstone peaks typically have dwarf-shrublands in which *Artemisia monosperma* and *Echiochilon fruticosum* are dominant (association of *Echiochilon fruticosum, Helianthemum sessiliflorum* and *Rhanterium garcinii* Long 1957, Artemision monospermae Eig 1946). The alluvial terraces of sandy loam at the foot of the sand hills are occupied by *Zilla spinosa* (*Anabasis articulata–Zilla spinosa* assoc. Feinbrun/Zohary 1955, Long 1957).

Floristic composition:

Sample record: N. of al-Quweira; alt. 1000 m; whitish sands of weathered Ordovician sandstones; exp. SE.; cov. up to 20% date: 13 Apr. 1980.

Shrubs: *Rhamnus palaestinus, Zilla spinosa.*

Chamaephytes: *Artemisia monosperma, Echiochilon fruticosum* var. *sieberi, Helianthemum lippii, H. sancti-antonii, Heliotropium digynum, Pituranthos triradiatus, Polygala sinaica, Scorzonera tortuosissima, Solanum sinaicum.*

Hemicryptophytes: *Anabasis syriaca, Astragalus aaronsohnianus, A. zemeraniensis, Centaurea eryngioides, Cyperus capitatus, Dianthus sinaicus, Psoralea flaccida, Salsola incanescens, Scrophularia xylorrhiza, Stipa parviflora, Stipagrostis ciliata.*

Therophytes: *Asteropterus leyseroides, Hippocrepis bicontorta, Linaria haelava, Matthiola aspera, Ononis serrata, Scabiosa porphyroneura.*

4.4 Vegetation units of the Nubo-Sindian type (Photo 9)

These are exclaves of the Sudanian region in Wādī l-ʿAraba and in the Jordan Valley. They contain many xero-tropical, African descendants whose physiognomy is reminiscent of the African thornbush savannas.

Thorn woodlands **(*5.2*, Photo 10)**

Bioclimate: arid, warm to very warm.

$8.8 < m < 12.0$ $2.7 < Q < 6$ $38 < P < 85$

Geology: On talus consisting of weathered Precambrian granites and diorites and Quaternary undifferentiated fluviatile gravel, talus and fans.

Distribution: Nubo-Sindian-influenced woody-plant formations form extensive, extremely xeromorphic woodlands resembling pseudo-savanna especially in the southern part of Wādī l-ʿAraba and around al-ʿAraba (Photo 10). This formation, which is dependent on groundwater and inflowing water, is confined to the runnels and wadis which cut the hammada plains, alluvial plains and alluvial fans. It contains xero-tropical African taxa which can be grouped sociologically with the Acacietum tortilidis anabasidetosum EIG 1946, in which there are character species of the Nubo-Sindian Acacietea and the Saharo-Arabian Anabasidetea. In the upper region of Wādī l-ʿAraba one also finds *Ziziphus spina-christi*, another component of the African thorn-bush savanna communities. In local exclaves there occur the following representatives, which here reach their northern limit of distribution: *Moringa peregrina* (association of *Ziziphus spina-christi–Moringa aperta* EIG 1946, especially in the area of the hot springs of Zarqāʾ Māʿīn), *Hyphaene thebaica* (character species of the xero-tropical wadi-margin communities, near al-ʿAqaba) and *Balanites aegyptiaca* (association of *Ziziphus spina-christi–Balanites aegyptiaca* EIG 1946, occuring north as far as Arīḫā [Jericho]).

Floristic composition:

Sample record: S. of al-ʿAqaba; alt. 50 m; on Quaternary talus and fans; exp. SW.; cov. 10%; date: 12 Apr. 1980.

Trees and shrubs: *Acacia raddiana, A. tortilis, Ochradenus baccatus, Pergularia tomentosa.*
Chamaephytes: *Cleome droserifolia, Hammada salicornica, Iphiona scabra.*
Hemicryptophytes: *Diplotaxis harra, Fagonia bruguieri, F. mollis, Forsskaolea tenacissima, Francoeria crispa, Heliotropium arbainense, H. bacciferum, Kickxia spartioides, Launaea mucronata, Morettia parviflora, Panicum turgidum, Polycarpaea repens, Stipagrostis ciliata.*
Therophytes: *Aizoon canariense, Erodium bryoniifolium, Ifloga spicata, Lotononis platycarpa, Schismus barbatus.*
Geophytes: *Asphodelus tenuifolius.*

Salt desert (18.1)

Bioclimate: arid, warm to very warm.
$8.8 < m < 12.0$ $2.7 < Q < 6$ $38 < P < 85$
Geology: Marls, clays and gypsum with clastic intercalations ("Lisan Marls").
Distribution: Covers large areas only on the Lisan peninsula and at the southern end (Ġaur) of the Dead Sea (al-Baḥr al-Mayyit). Owing to the high content of water-soluble salts in these white-grey marls and to the aridity of this substrate, the sparse vegetation is restricted to a few channels.

Cultivated land (21.)

Distribution: In Wādī l-ʿAraba there are several large oases (Ġaur aṣ-Ṣāfī) which are characterized by alluvial, non-saline, moist soils. Here one mainly finds shrub formations consisting of *Moringa peregrina* and *Salvadora persica.* In the oases the land is used for cultivation of dates, sugar cane, millet and vegetables (mainly *Irrigated arable land*).

4.5 Review of the vegetation

In conclusion Fig. 2 and Table 2 (a–d) present a survey of the characteristic formations of southern Jordan, showing their dependence on a complex of edaphic and bioclimatic factors.
The description and documentation of these formations makes it possible to classify the natural regions on a phytocoenological basis. By means of this classification a description can be given of successions and of changes in the potential of the natural region brought about by man and animals. This lays the groundwork not only for further ecophysiological studies on the effects of various factors but also for protection and recultivation measures in this area, which will become increasingly necessary in the future.

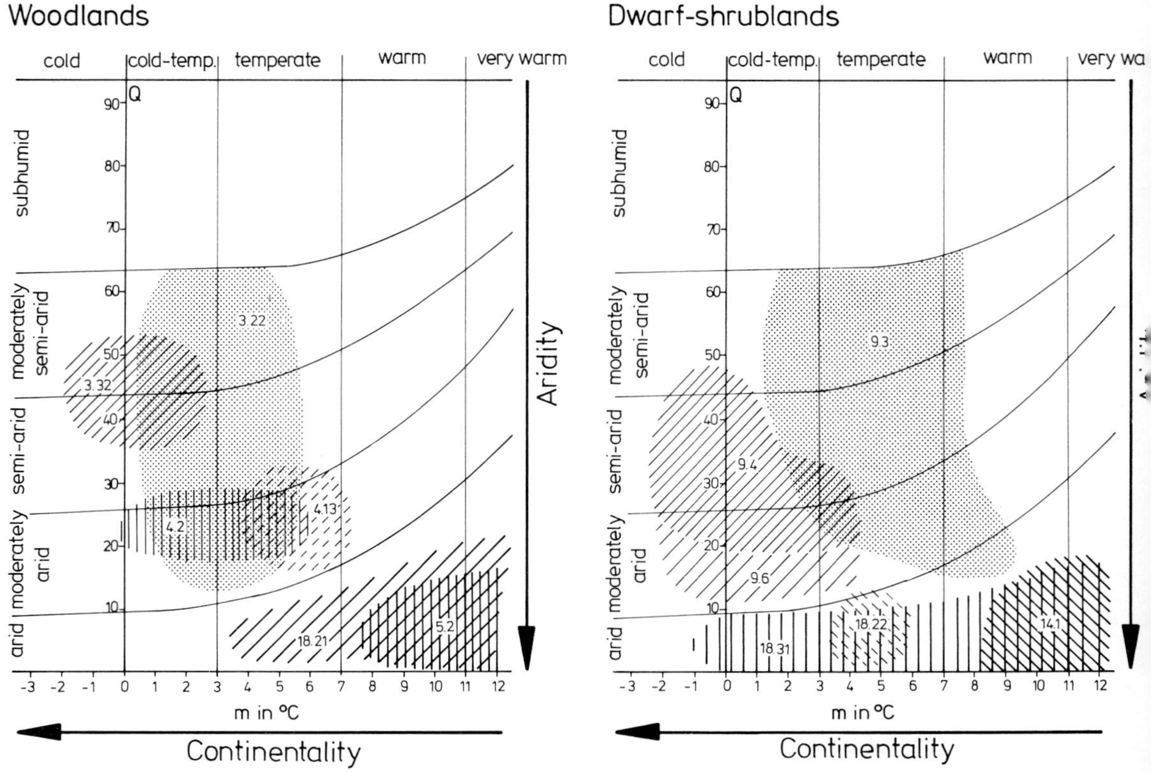

- 3.22 Evergreen broad-leaved woodland relatively resistant to cold
- 3.32 Evergreen needle-leaved woodland resistant to cold
- 4.13 Cold-deciduous montane woodland with evergreen needle-leaved trees or shrubs
- 4.2 Cold-deciduous broad-leaved woodlands without evergreens
- 5.2 Thorn woodlands
- 18.21 Sand desert with shrubs

- 9.3 Mixed dwarf-scrub and herbaceous formations
- 9.4 Xeromorphic dwarf-shrublands
- 9.6 Xeromorphic, very open dwarf-shrublands
- 14.1 Halophytic semi-woody shrub communities
- 18.22 Sand desert with dwarf-shrubs
- 18.31 Rock desert

Fig. 2 Bioclimatogram of the characteristic vegetation units in southern Jordan (cf. also Fig. 1).

	Vegetation unit	Bioclimate	Geology	Distribution
Woodlands	*Evergreen broad-leaved woodland relatively resistant to cold (Quercus calliprinos* community type)	Moderately arid to semiarid, cold-temperate to temperate	Cenomanian limestones and dolomites	Western border mountains, 1200–1600 m
Woodlands	*Evergreen needle-leaved woodland resistant to cold (Juniperus phoenicea* community type)	Semi-arid, cold	Cambrian massive sandstones and Precambrian granites	Western border mountains, (600) 900–1500 m
Woodlands	Fragments of *Cold-deciduous montane woodland with evergreen needle-leaved trees or shrubs* (*Pistacia khinjuk – Ceratonia siliqua* community type)	Moderately arid, temperate	Precambrian granites and Cambrian massive sandstones	Western border mountains, 600–900 m, in gorges and crevices
Dwarf-shrublands	*Mixed dwarf-scrub and herbaceous formations*			
Dwarf-shrublands	– *Sarcopoterium spinosum* community type	Semi-arid to subhumid, temperate	Cenomanian limestones (terra rossa)	Western hills and plateaus (Moab, N. Edom)
Dwarf-shrublands	– *Salvia dominica – Ballota undulata* community type		Turonian limestones (greyish soils)	

Table 2 a. Vegetation units of the Mediterranean type (overview).

	Vegetation unit	Bioclimata	Geology	Distribution
Woodlands	*Cold-deciduous broad-leaved woodlands without evergreens* (*Pistacia atlantica* community type)	Moderately arid, cold-temperate to temperate	Cenomanian and Santonian-Turonian limestones	Western border mountains and plateaus, 1400 m
Dwarf-shrublands	*Xeromorphic dwarf-shrublands*			
Dwarf-shrublands	– *Artemisia herba-alba* community type	Moderately arid, cold-temperate to temperate	Santonian-Turonian limestones (greyish soils)	Western hills and plateaus (Edom), 800–1500 m
Dwarf-shrublands	– *Haloxylon articulatum* community type		Loess soils in gullies and runnels	

Table 2 b. Vegetation units of the Irano-Turanian type (overview).

	Vegetation	Bioclimate	Geology	Distribution
Shrublands	*Sand desert with shrubs*			
	– *Haloxylon persicum* community type	Arid, temperate to warm	Eolian sands, fine-sandstones and sandy shales	SE. desert (surrounding of al-Mudawwara) and Wādī l-'Araba
	– *Hammada salicornica* community type	Arid, temperate	Eolian sands, weathered Cambrian sandstones	S. desert (surrounding of Wādī Ramm)
Dwarf-shrublands	*Sand desert with dwarf-shrubs* – *Echiochilon fruticosum* community type – *Zilla spinosa* community type	Arid, temperate	Massive, whitish weathered Ordovician sandstones	Mainly between al-Quwaira and Ra's an-Naqb
	Rock desert (*Anabasis articulata* community type)	Arid, cold-temperate to warm	Conglomeratic limestones, Quaternary mantle rocks	E. desert, locally in Wādī l-'Araba
	Debris and gravel desert (litho- and chasmophytic associations)	Arid, cold-temperate to temperate	Precambrian granites, Cambrian sandstones	Isolated mountain ridges in the south
	Halophytic semi-woody shrub communities	Arid, warm to very warm	Marls, clays and gypsum with clastic intercalations	On azonale hydro- and subhydrohalopedic soils (mainly Wādī l-'Araba)

Table 2c. Vegetation units of the Saharo-Arabian type (overview).

	Vegetation unit	Bioclimate	Geology	Distribution
Woodlands	*Thorn woodlands* (*Acacia tortilis* community type)	Arid, warm to very warm	Talus, fans and undifferentiated fluviatile gravel	Southern parts of Wādī l-'Araba and surroundings of al-'Aqaba

Table 2d. Vegetation units of the Nubo-Sindian type (overview).

5. References

Al-Eisawi, D. M., List of Jordan vascular plants, Mitt. Bot. Staatssamml. München *18*, 1982, 79–182.

Al-Eisawi, D. M., Studies on the flora of Jordan 10. Nine new species to the flora of Jordan, Candollea *38*, 1983 a, 359–364.

Al-Eisawi, D. M., Studies on the flora of Jordan 11. On the flora of Wadi Araba (Araba Valley), Candollea *38*, 1983 b, 365–385.

Atkinson, K. / Beaumont, P., The forests of Jordan, Economic Botany, *25*, 1971, 305–311.

Bardtke, H., Die Waldgebiete des jordanischen Staates, Zeitschr. dtsch. Palästina Ver. *72*, 1956, 109–122.

Bender, F. et al., Geological map of Jordan, sheet Aqaba-Ma'an 1 : 250 000, Bundesanstalt f. Bodenforschung, 1968, Hannover.

Boulos, L., Studies on the flora of Jordan 5. On the flora of El-Jafr–Bayir Desert, Candollea *32*, 1977, 99–109.

Boulos, L. / Al-Eisawi, D. M., Studies in the flora of Jordan 6. On the flora of Ras en Naqb, Candollea *32*, 1977 a, 111–119.

Boulos, L. / Al-Eisawi, D. M., Studies on the flora of Jordan 8. New and noteworthy plants, Candollea *32*, 1977 b, 270–276.

Boulos, L. / Jallad, W., Studies on the flora of Jordan 1. *Diplotaxis villosa* sp. nov. (Cruciferae), Bot. Notiser *128*, 1975, 365–367.

Boulos, L. / Jallad, W. / Lahham, J., Studies on the flora of Jordan 2. Seven species new to the flora of Jordan, Bot. Notiser *128*, 1975, 368–370.

Boulos, L. / Lahham, J., Studies on the flora of Jordan 3. On the flora of the vicinity of the Aqaba gulf, Candollea *32*, 1977 a, 73–80.

Boulos, L. / Lahham, J., Studies on the flora of Jordan 4. On the desert flora northeast of Aqaba, Candollea *32*, 1977 b, 81–98.

Boulos, L. / Lahham, J., Studies on the flora of Jordan 9. *Kickxia azraqensis* (Scrophulariaceae), sp. nov., Candollea *32*, 1977 c, 277–280.

Boulos, L. / Lahham, J. / Jallad, W., Studies on the flora of Jordan 7. On the desert flora of the area of H 4 and H 5 pumping stations, Candollea *32*, 1977, 255–268.

Chapman, J. D., Notes on the occurrence of *Cupressus sempervirens* in Transjordan, Palest. J. Bot., Jerusalem Ser. *4*, 1947, 55.

Dubieff, J., Le climat du Sahara, Mém. l'Inst. Rech. Saharienne, Univ. Alger., 1959.

Eig, A., Les éléments et les groupes phytogéographiques auxiliaires dans la flore palestinienne, Feddes Rep. Beih. *63*, 1931, 1–201.

Eig, A., On the phytogeographical subdivision of Palestine, Palest. J. Bot., Jerusalem Ser. *1*, 1938, 4–12.

Eig, A., Synopsis of the phytosociological units of Palestine, Palest. J. Bot., Jerusalem Ser. *3*, 1946, 183–246.

Emberger, L., Sur une formule climatique et ses applicables en géographie botanique, Compt. rend. hebd. Sci. Acad. Science *191*, 1930, 389–390.

Emberger, L., Aperçu général sur la végétation du Maroc, Veröff. Geobot. Inst. Rübel *14*, 1939, 40.

Emberger, L., Sur le quotient pluviothermique, Compt. rend. hebd. Sci. Acad. Science *234*, 1952, 2508–2510.

EVENARY, M. / ORSHANSKY, G., The Middle Eastern hammadas, Lloydia *11*, 1948, 1–13.

FEINBRUN, N., Flora Palaestina Vol. *3*, 1978, Jerusalem.

FEINBRUN, N. / ZOHARY, M., A geobotanical survey of Transjordan, Bull. Res. Counc. Israel Sect. D *5*, 1955, 5–36.

FRANKENBERG, P., Lebensformen und Florenelemente im nordafrikanischen Trokkenraum, Vegetatio *37*, 1978, 91–100.

FREY, W. / KÜRSCHNER, H., Photosyntheseweg und Zonierung von Halophyten an Salzseen in der Türkei, in Jordanien und im Iran, Flora *173*, 293–310.

FREY, W. / PROBST, W., Gliederung der Vegetation und ihre Darstellung im Tübinger Atlas des Vorderen Orients (TAVO) und in den Beiheften zum Atlas, Beih. Tübinger Atlas Vorderer Orient, Reihe A, *1*, 1977, Wiesbaden.

KASAPLIGIL, B., A report on the vegetation profiles of the forest and grazing lands in Jordan, FAO 1956, Amman.

LAWRENCE, T. E., Seven pillars of wisdom, 1936, London.

LONG, G. A., The bioclimatology and vegetation of Eastern Jordan, FAO Report, Rome 1957, FAO/57/2/1109.

MONOD, TH., Modes "contracté" et "diffus" de la végétation saharienne, in: CLOUDSLEY-THOMPSON, J. L. (ed.), Biology of deserts, Proc. Symp. Biol. of hot and cold deserts, London, 1954, 35–44.

ZOHARY, M., Outline of the vegetation in Wadi Araba, J. Ecol. *32*, 1945, 204–213.

ZOHARY, M., Flora Palaestina Vol. *1*, 1966, Vol. *2*, 1972, Jerusalem.

ZOHARY, M., Geobotanical foundations of the Middle East, 1973, Stuttgart.

ZOHARY, M. / FEINBRUN, N., Outline of the vegetation of the Northern Negev, Palest. J. Bot. Jerusalem Ser. *5*, 1951, 96–114.

ZOHARY, M. / ORSHAN, G., Ecological studies in the vegetation of the Near East deserts II. Wadi Araba, Vegetatio *7*, 1956/57, 15–37.

ZOHARY, M. / ORSHANSKY, G., Structure and ecology of the vegetation in the Dead Sea region of Palestine, Palest. J. Bot. Jerusalem Ser. *4*, 1949, 177–206.

Dr. Harald KÜRSCHNER, Institut für Systematische Botanik und Pflanzengeographie, Freie Universität Berlin, Altensteinstr. 6, D-1000 Berlin 33.

Photo 1 *Evergreen broad-leaved woodland relatively resistant to cold* (*Quercus calliprinos* woodland) N. of Baṭrā.

Photo 2 *Evergreen needle-leaved woodland resistant to cold (Juniperus phoenicea* woodland).

Photo 3 Remnants of *Cold deciduous broad-leaved woodlands without evergreens (Pistacia atlantica)* near aš-Šaubak.

Photo 4 *Xeromorphic dwarf-shrublands* near aš-Šaubak (Artemisietum herbae-alba).

Photo 5 *Sand desert with shrubs* (Hammadetea salicornici). Wādī Ramm.

Photo 6 *Sand desert with shrubs* (Haloxylonetum persici). Wādī l-ʿAraba.

Photo 7 *Rock desert* (degradated Anabasidetea articulatae) E. of Maʿān.

Photo 8 *Debris and gravel desert.* Ǧabal Ramm.

Photo 9 Vegetation units of Nubo-Sindian type.

Photo 10 *Thorn woodlands* (Acacietum tortilidis anabasidetosum) N. of al-'Aqaba.

Riad Halwagy

On the ecology and vegetation of Kuwait

Contents

1. Introduction

The State of Kuwait (28°30'–30°05'N, 46°33'–48°30'E lies in the northern part of the Arabian Peninsula. The land is flat to gently undulating, rising gradually from the shores of the Arabian Gulf in a southwesterly direction and reaching a maximum height of 295 m (Fig. 1). The flat open desert is interrupted by five main geomorphologic features: ridges, wadis (dry drainage lines), depressions, sand dunes and salt marshes (Milton 1967; Fuchs et al. 1968; Halwagy/Halwagy 1974 a).

2. Soils

The soils of Kuwait had been formed under the prevailing arid climate and generally show weak profile development. Four great soil groups could be distinguished (Ergun 1969; Halwagy/Halwagy 1974 a).

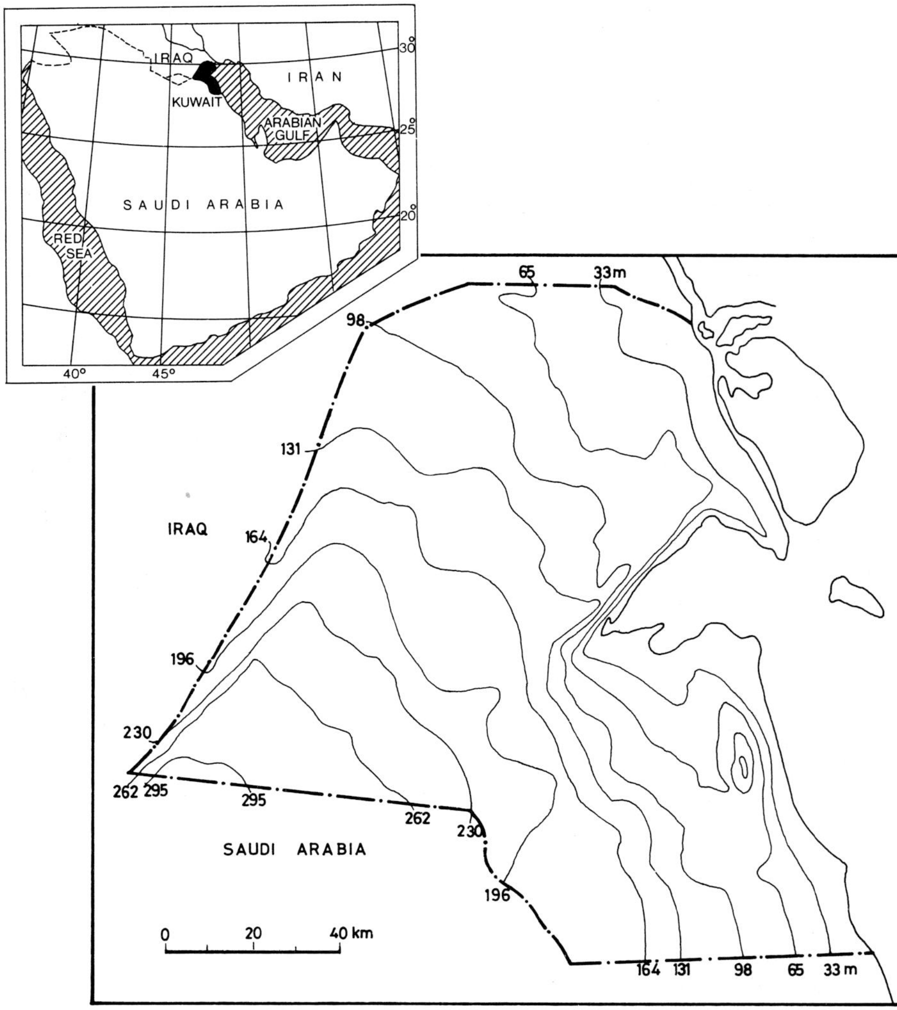

Fig. 1 Location map and contour map of Kuwait (ERGUN 1969).

A. Desert soils

These soils comprise the main bulk of soils of the country (Fig. 2). They have superficial deposits (0–150 cm deep) of coarse sands, loamy sands or sandy loams over a hardpan locally known as "gatch". The hardpan varies from place to place and is generally calcareous and/or gypseous and/or gravelly and/or saline. The soils become shallower toward the north and west of the country, the hardpan frequently outcropping on the surface. Surface soils have single grain structure and good to imperfect drainage. According to the Comprehensive System

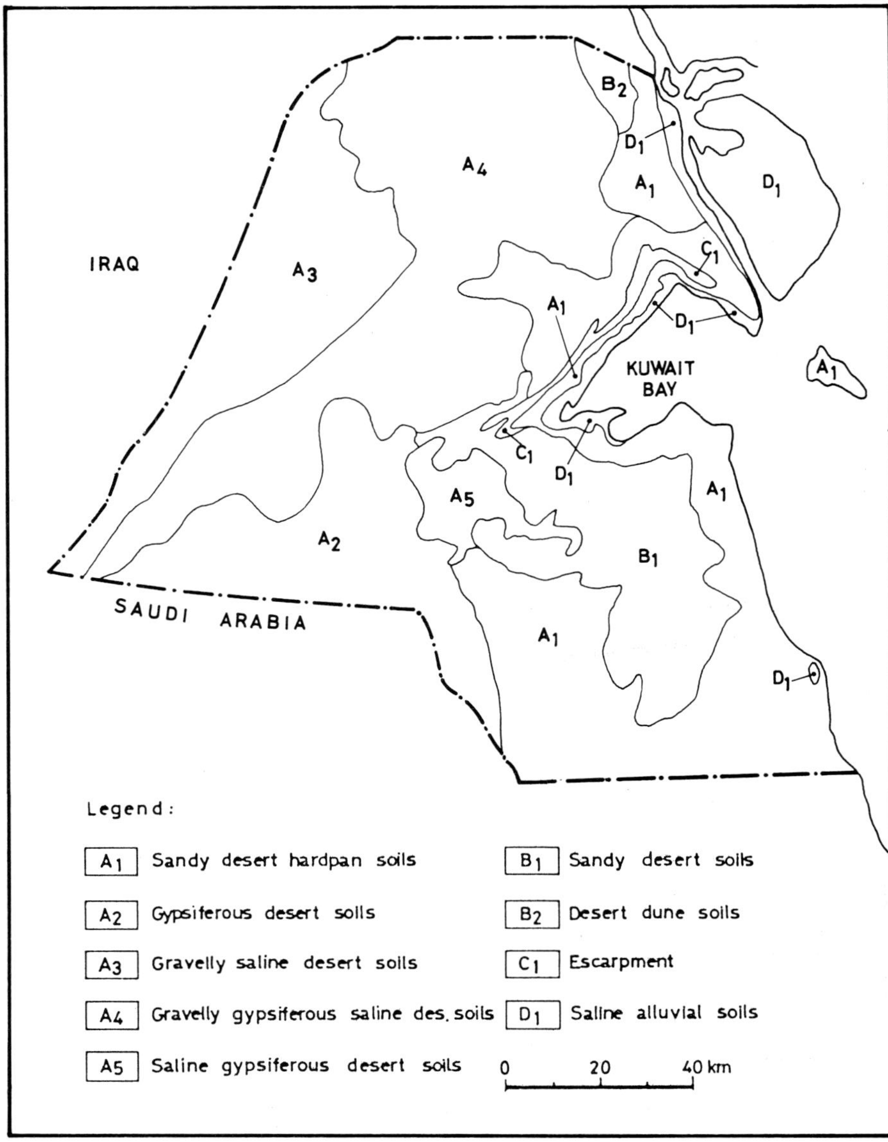

Fig. 2 Soil map of Kuwait (modified after ERGUN 1969).

(USDA 1960), these soils belong to soil Order Aridisols, Suborder Orthids.

B. Desert-regosol intergrade soils

These occur south of Kuwait City and in the extreme northeast of the country. They are deep (more than 200 cm) soft soils, mostly of coarse sands or sandy loams which are fairly easily drifted. They have single grain structure and good to excessive drainage. They correspond to soil Order Entisols, Suborder Psamments of the Comprehensive System.

C. Lithosols

Lithosols are restricted to Jal-Az-Zor escarpment which borders the northern shores of the Bay of Kuwait (Fig. 3). They are shallow soils (0–50 cm), of sandy to gravelly loamy texture overlying the sedimentary and gypsiferous sandstone of the escarpment. They have good to excessive drainage and rapid run off. They are equivalent to Order Entisols, Suborder Orthents.

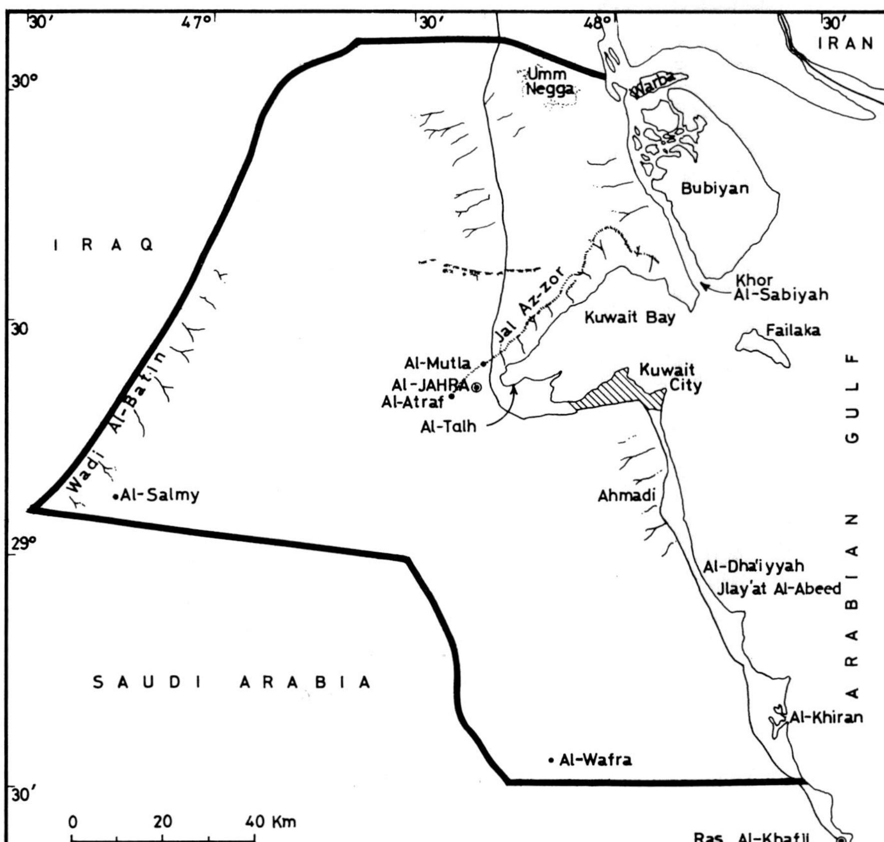

Fig. 3 District map of Kuwait.

D. Saline alluvial soils

These soils occur along the coast and consist of soft deposits which are often calcareous and/or gypseous. They are covered by deltaic and tidal sediments which range in texture from shelly sand and sandy loam to sandy clay and clay. Drainage may be good to imperfect. Platy structure often develops on the surface.

3. Climate

Hot dry summers and cool rainy winters characterise the climate of Kuwait. The map of the world distribution of arid regions (UNESCO 1979) desingates it as "arid, with mild winter, very warm summer, dominant summer drought, winter precipitation regime, and P/Etp (annual precipitation/potential evapotranspiration) 0.03–0.20".
The main climatic features are illustrated (Fig. 4) using the notation of Walter (1973). This figure is more accurate than that given by Walter et al. (1975) in their "Klimadiagramm-Karten" for Kuwait which is based on

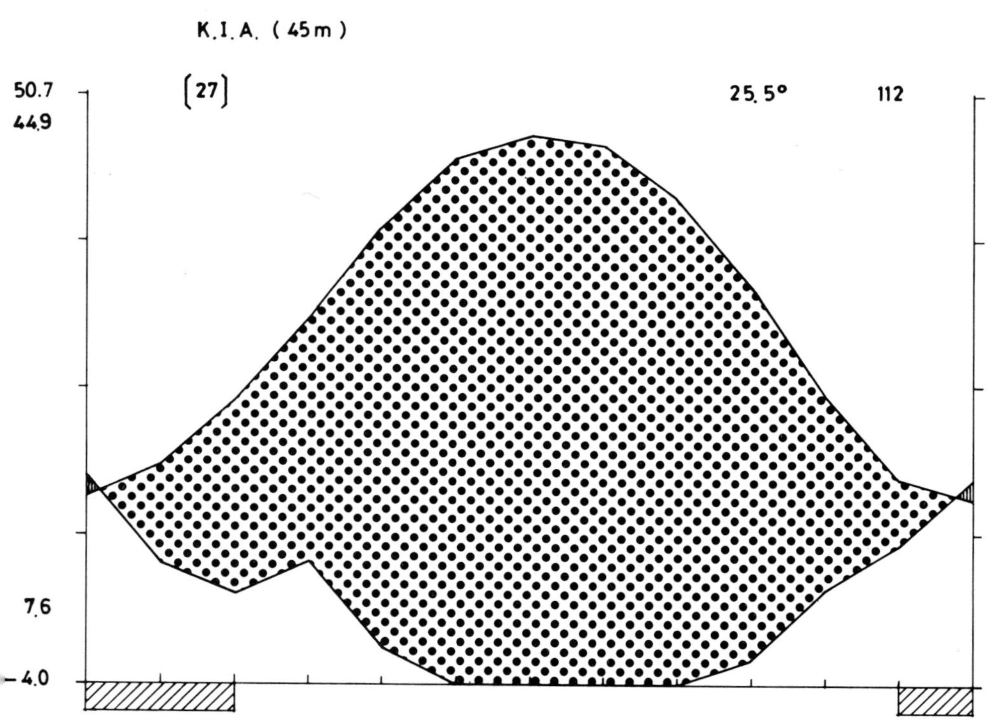

Fig. 4 Climatic characteristics of Kuwait using the notation of Walter (1973). Data measured at Kuwait International Airport (K.I.A.) 1957–1983.

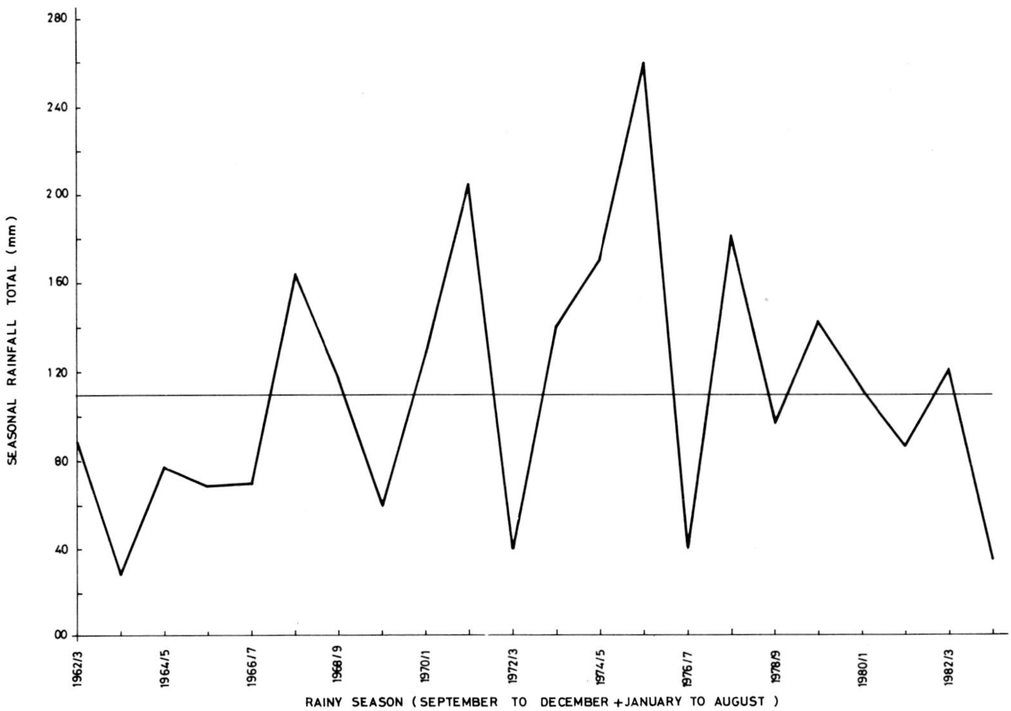

Fig. 5 Rainfall variability.

7 years only. From Fig. 4 it is clear that, with the exception of January, all months are arid, with potential evapotranspiration exceeding rainfall. The mean seasonal rainfall (October–May) is 112 mm, with great variations from year to year (Fig. 5) and from one place to another. Extremes on record are 28.1 mm (1963/64) and 260.2 mm (1975/76). The rate of Piche evaporation is highest in July (32.2 ml/day) and lowest in January (4.8 ml/day). The prevailing winds blow from the northwest and bring hot air in summer and cool or cold air in winter. They frequently raise dust and sand and may cause dust and sandstorms. Southeastern winds are less frequent and bring humid air from the sea.

4. Vegetation

4.1 General features

The vegetation of Kuwait is a poor open scrub of undershrubs, perennial herbs and grasses and short spring ephemerals. On favoured sites, however, we may encounter hardy shrubs about man's height. The desert vegetation is primarily a function of the prevailing climate, particularly the meagre rainfall. This may be modified by landform and biotic factors. The

density and height of the vegetation cover bear a direct relationship to the total annual rainfall. In years of above average rainfall, e.g. 1971/72 and 1975/76, with seasonal total of 204 and 260 mm respectively, the desert becomes covered with a rich green mantle. On the other hand, years of low rainfall, e.g. 1962/63 (28.1 mm) and 1983/84 (34.7 mm) witness almost complete failture of vegetation. The length of the rainy season determines the length of the growth season. The first rains trigger the germination of annuals and stimulate the sprouting or perennials. Good growth is ensured if the first rains fall during the "warm" period (October 10–November 5) when warm weather still prevails and favours seed germination and bud development. When the first rains are delayed to December or January, they may coincide with cold temperatures and do not benefit plant activity. The end of the rainy season (often in April) brings about the maturation and drying up of the desert vegetation. A shortened or prolonged rainy season causes early or late cessation of plant activity.
Landform, particularly microrelief, modifies the effect of rainfall through influencing the redistribution of water and sediments, directing them to lower ground. Such favoured sites, e.g. drainage lines and sandy basins usually support a much more luxuriant vegetation cover than adjacent areas which are only a few decimetres higher up. The situation may be further ameliorated when this lush vegetation traps drifted sand and builds phytogenic mounds. The seasonality of climate (winter rain) determines the seasonal aspect of the vegetation: a relatively rich green vernal aspect alternates with a dry, poor almost lifeless aestival and autumnal aspect. This contrast becomes more marked due to overgrazing by sheep, goats and camels which help to denude the desert of its thin plant cover, particularly during the long dry season. Omar et al. (1981) estimated that animals grazing the natural vegetation during March–April 1980 were: 540980 sheep, 94494 camels, 16881 goats, 586 horses, 410 donkeys and 104 cows.

4.2 Successional relationships

Botanical studies in Kuwait are very recent. The earliest vegetation map dates back to 1955 (Dickson 1955). This was followed by several maps (Kernick 1964; Kernick 1966; Ergun 1969; Macksad 1969). Halwagy/Halwagy (1974b) studied these maps in detail in an attempt to understand the possible successional relationships and also proposed a vegetation map as in 1974 (Fig. 6). They suggested the following sequence of succession:

- Communities of *Cyperus conglomeratus* and *Rhanterium epapposum* replace that of *Panicum turgidum* under the prevailing conditions of overgrazing since *Panicum* is more palatable.

– The community of *Hammada salicornica* expands at the expense of that of *Rhanterium epapposum* as a result of overgrazing, shrub cutting or soil erosion. *Rhanterium* is more palatable and was more in demand as a source of fuel before the discovery of oil. Moreover, *Hammada* appears to survive on thin soil (10–50 cm) above hardpan, whereas *Rhanterium* usually flourishes on deeper soil.

4.3 Vegetation types

The vegetation can be described under four ecosystems (HALWAGY/HALWAGY 1974b) which are distinguished on the basis of habitat factors (soil charateristics and landform) and floristic composition (dominant species). These ecosystems are: The sand dune ecosystem, the salt marsh and saline

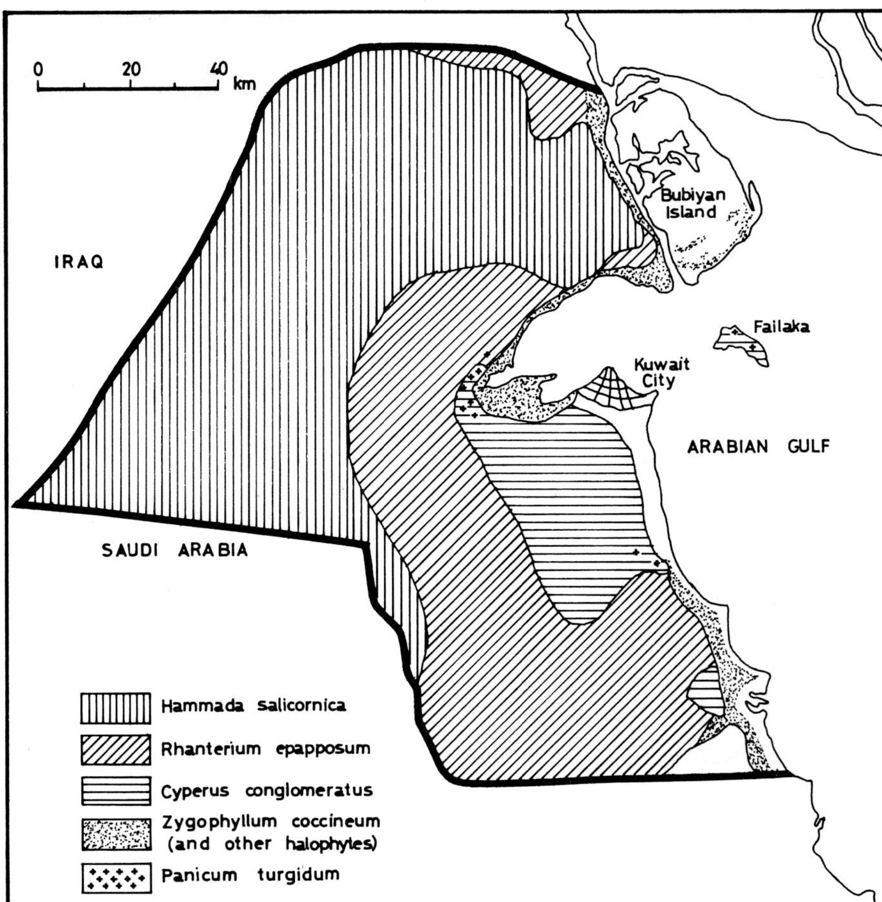

Fig. 6 Vegetation map of Kuwait (HALWAGY/HALWAGY 1974b).

depressions ecosystem, the desert plain ecosystem and the desert plateau ecosystem. The first two ecosystems show marked differences in vegetation and soil, while the desert plain grades imperceptibly into the desert plateau, with obvious differences in the dominant species and only subtle differences in soil type.

4.3.1 The sand dune ecosystem

This ecosystem consists essentially of a series of low dunes which extend along the Gulf coast from Al-Dba'iyyah southwards. It forms part of an oolitic complex of Pleistocene origin in which Picha/Salfh (1975) distinguished five mappable lithostratigraphic units: Recent oolitic sands and beach rocks, ridges of younger oolitic limestones, ridges of older oolitic limestones, ridges of quartz-oolitic sandstones, and oolitic quartzose sandstones. These ridges vary in height from 5 to 15 metres and separate the open sea from the coastal lagoons and "sabkhas" (supratidal saline flats) of variable width. The ridges are partly marine and partly aeolian and consist of more or less consolidated sediments which are partly covered by recent dunes and beaches. The superficial soil thus varies from loose wind-blown predominantly oolitic coarse sand to slightly or completely cemented material.
The sand dune ecosystem shows considerable local variations from one site to another. However, we can distinguish at least 5 habitat types: Embryo dunes, seaward face of the dune ridge, crest of the dune ridge, leeward slope of the dune rigde, and dune foot. They differ mainly in the degree of protection from wind, mobility of the substratum and nature of the surface sediments. The floristic composition of a representative site at Jlay'at Al-Abeed is shown in table 1.
The salient features of the 5 habitat and vegetation types of the sand dune ecosystem may be summarized as follows:

Embryo dunes

These occur along the foreshore beyond the reach of the highest spring tides and generally vary from 1 to 3 metres in height and show various degrees of coalescence. The oolitic sand is very mobile and is usually colonized by a sole dominant (*Seidlitzia rosmarinus* or *Halopyrum mucronatum*). Only well adapted species can withstand this difficult habitat. Total coverage is very low, hardly reaching 5%.

Seaward face of the dune ridge

The main dune ridge rises for 5–15 metres and the seaward face consists of deep loose oolitic sand, with or without consolidated outcrops. Dominant species are *Zygophyllum coccineum* or *Atriplex leucoclada*, with *Pennisetum divisum*, *Aellenia glauca*, *Seidlitzia rosmarinus*, *Halopyrum mucrona-*

Species	ED	SF	C	LS	F
Perennials					
Halopyrum mucronatum (L.) Stapf	d	o	–	–	–
Atriplex leucoclada Boiss.	–	d	cod	c	o
Aellenia glauca (M. Bieb.) Aellen	–	c–a	cod	r	r
Launaea nudicaulis (L.) Hook. f.	–	o–c	–	–	–
Salsola baryosma (Schult.) Dandy	–	o	r	o	r
Cynodon dactylon (L.) Pers.	–	o	–	–	–
Panicum turgidum Forssk.	–	o	o	o	–
Pennisetum divisum (Gmel.) Henr.	–	r–o	r	o	–
Lycium shawii Roem. et Schult.	–	r	r	–	–
Cistanche tubulosa (Schenk) Wight	–	r	–	–	r
Heliotropium bacciferum Forssk.	–	r	–	–	r
Rhanterium epapposum Oliv.	–	r	r	r	–
Fagonia olivieri DC.	–	–	o	r	r
Helianthemum lippii (L.) Pers.	–	–	r	r	r
Anabasis setifera Moq.-Tand.	–	–	r	r	–
Convolvulus buschiricus Bornm.	–	–	r	–	r
Haplophyllum tuberculatum (Forssk.) Adr.-Juss.	–	–	r	r	o
Onobrychis ptolemaica (Del.) DC.	–	–	r	–	–
Orobanche cernua Loefl.	–	–	r	–	–
Launaea mucronata (Forssk.) Muschl.	–	–	r	r	r
Zygophyllum coccineum L.	–	–	r	d	d
Sporobolus arabicus Boiss.	–	–	r	o	r–o
Stipagrostis plumosa (L.) Munro ex T. Anders.	–	–	–	r	–
Seidlitzia rosmarinus (Ehrenb.) Solms.-Laub.	–	–	–	r	–
Hammada salicornica (Moq.) Iljin	–	–	–	r	–
Herniaria hemistemon J. Gay	–	–	–	r	o–c
Aeluropus sp.	–	–	–	–	r
Biennials and annuals					
Launaea capitata (Spreng.) Dandy	–	o	o–c	o	r
Cakile arabica Velen.	–	o	o	r	–
Schismus barbatus (L.) Thell.	–	o	c–a	c	–
Salsola jordanicola Eig	–	o	o–c	c	o
Erodium laciniatum (Cav.) Willd.	–	r	o	o	–
Avena fatua L.	–	r	–	–	–
Lotus halophilus Boiss. et Sprun.	–	r	c	o	–
Senecio desfontainei Druce	–	r	c	o	o
Lolium temulentum L.	–	r	r	–	–
Phalaris minor Retz.	–	–	o	–	–
Atractylis carduus (Forssk.) Christ.	–	–	o	o	r
Medicago laciniata (L.) Mill.	–	–	o	o	o

Tab. 1 Floristic composition of a representative east-west transect of the coastal sand dune system at Jlay'at Al-Abeed (ED = embryo dunes, SF = seaward face of main ridge, C = crest of main ridge, LS = leeward slope of main ridge, F = foot of main ridge; d = dominant, cod = codominant, a = abundant, c = common, o = occasional, r = rare).

Biennials and annuals/contd.

Species	ED	SF	C	LS	F
Paronychia arabica (L.) Del.	–	–	r	r	–
Neurada procumbens L.	–	–	r	r	–
Calendula aegyptiaca Desf.	–	–	r	r	–
Trigonella stellata Forssk.	–	–	r	r	r
Emex spinosa L.	–	–	r	–	r
Koelpinia linearis Pall.	–	–	r	–	–
Silene arabica Boiss.	–	–	r	–	–
Bienertia cycloptera Bge.	–	–	–	o	a
Gastrocotyle hispida (Forssk.) Bge.	–	–	r	r	–
Cutandia memphitica (Spreng.) Benth.	–	–	r	–	–
Hordeum glaucum Steud.	–	–	r	–	–
Gypsophila antari Post	–	–	r	r	r
Chenopodium murale L.	–	–	r	–	–
Arnebia decumbens (Vent.) Coss. et Kral.	–	–	r	–	r
Hippocrepis bicontorta Lois.	–	–	r	r	r
Malva parviflora L.	–	–	r	–	–
Astragalus annularis Forssk.	–	–	r	o	–
Ifloga spicata (Forssk.) Sch.-Bip.	–	–	r	c	–
Asphodelus fistulosus L. var. *tenuifolius* Cav.	–	–	r	–	–
Spergularia diandra (Guss.) Heldr. et Sart.	–	–	–	o	c
Mesembryanthemum nodiflorum L.	–	–	–	o	c
Anthemis deserti Boiss.	–	–	–	r–o	r
Oligomeris linifolia (Vahl) Macbride	–	–	–	r	r
Reichardia tingitana (L.) Roth.	–	–	–	r	–
Plantago boissieri Hausskn. et Bornm.	–	–	–	r	–
Plantago ovata Forssk.	–	–	–	r	–
Astragalus corrugatus Link	–	–	–	r	r
Anisosciadium lanatum Boiss.	–	–	–	r	–
Stipa capensis Thunb.	–	–	–	r	–
Sphenopus divaricatus (Gouan) Reichenb.	–	–	–	–	o
Frankenia pulverulenta L.	–	–	–	–	r
Astragalus tribuloides Del.	–	–	–	–	r
Plantago coronopus L.	–	–	–	–	r
Filago desertorum Pomel	–	–	–	–	r
Allium desertorum Forssk.	–	–	–	–	r
Gymnarrhena micrantha Desf .	–	–	–	–	r
Lophochloa pumila (Desf.) Bor					

tum, *Panicum turgidum* and *Cistanche tubulosa* as common associates. The latter species is a root parasite on *Seidlitzia* and *Zygophyllum*. Total cover is about 10% and the number of species may vary between 12 and 21.

Crest of the dune ridge

The crest usually has an undulating surface which may be loose or rather firm. Total plant coverage often ranges between 10 and 15% but may reach up to 35%. Species diversity is high (14–47 species). Dominance often goes to *Zygophyllum coccineum*, occasionally to *Seidlitzia rosmarinus* or *Atriplex leucoclada.* Common species include *Aellenia glauca*, *Pennisetum divisum* (Photo 1) and *Lycium shawii.*

Leeward slope of the dune ridge

The leeslope is fairly well protected from the effects of wind. This protection is reflected in soil conditions and plant life. The soil varies from rather loose mobile to fixed firm deposits. Denser plant cover (up to 40%) is usually associated with sites of moderately fixed soil. The total number of species may vary from 23 to 49. *Zygophyllum coccineum* is often the dominant species. Important species are *Seidlitzia rosmarinus*, *Aellenia glauca*, *Cistanche tubulosa*, *Cornulaca leucacantha*, *Salsola baryosma* and *Heliotropium bacciferum.*

Dune foot

This habitat is often characterized by a flat or very gently undulating firm surface of rather loamy fixed sand. Plant coverage ranges from 20 to 50%. Species diversity is high (38–47 species). *Zygophyllum coccineum* is again the dominant species. Common associates are: *Seidlitzia rosmarinus*, *Atriplex leucoclada*, *Salsola baryosma*, *Lycium shawii*, *Aellenia glauca*, *Pennisetum divisum*, and *Rhanterium epapposum.*

It should perhaps be added that when a dune overlies a saline bed, the vegetation usually includes a distinct halophytic element, e.g. *Nitraria retusa*, *Bienertia cycloptera*, *Frankenia pulverulenta*, *Spergularia diandra* and *Mesembryanthemum nodiflorum.*

To the west of the dune complex lies an extensive more or less flat sandy plain. The soil is predominantly of quartzose fixed sand which may occasionally be slightly saline and develop a weak crust. Plant cover varies between 20 and 40% and species range from 24 to 40. *Zygophyllum coccineum* dominates here as well, and important species included are *Seidlitzia rosmarinus*, *Salsola baryosma*, *Atriplex leucoclada*, *Lycium shawii*, *Aellenia glauca*, *Cornulaca leucacantha* and *Panicum turgidum.*

It thus appears that in this coastal dune habitat, increased shelter from the full effect of wind and decreased mobility of the substratum lead to an increase in total plant coverage and species diversity. The foreshore embryo dunes are often colonized by a sole dominant of highly adapted tall perennial species, but as the habitat conditions improve, less specialized species invade the dune, including a large number of small annuals. This may be, in a way, comparable to the well-known dune succession in temperate and colder more rainy climates of western Europe (see for example HALWAGY 1953; TANSLEY 1965). However, in Kuwait, succession is usually arrested at an early stage, possibly equivalent to the late white dune stage (Tansley 1965). This is due primarily to the paucity of rainfall, high rate of evaporation, and overgrazing which do not permit the development of a dense vegetation cover which could otherwise exert an influence on the habitant (autogenic element).

4.3.2 The salt marsh and saline depressions ecosystem

Salt marshes occur mainly along the coast of Kuwait Bay and Khor Al-Sabiyah and also on Bubiyan and Warba Islands. Smaller coastal marshes also occur along the Arabian Gulf in the southeastern part of the country. Saline depressions (sabkhas) usually occur on either side of the Kuwait-Ras Al-Khafji highway to the west of the coastal dunes.

Saline depressions vary in size and their soil ranges from loamy sand to sandy clay. They are influenced mainly by microtopography and the shallow saline water table. The centre of the depression may be barren or covered with *Halocnemum strobilaceum.* The water table is generally 60–80 cm below the surface. The *Halocnemum* zone is commonly fringed higher up by *Bienertia cycloptera.* The uppermost (sandy) edges support *Zygophyllum coccineum.*

The coastal salt marshes which fringe the Bay of Kuwait and Khor Al-Sabiyah have been extensively studied (Halwagy 1973; Halwagy/Halwagy 1977). They are influenced chiefly by tidal action (maximum tidal range 4.24 m, minimum 0.06 m), the saline water table and microrelief. They are occupied by 11 halophytic community types which are recognized chiefly by the dominant species. These include 5 major community types which cover extensive areas: *Halocnemum strobilaceum, Nitraria retusa, Tamarix macrocarpa, Seidlitzia rosmarinus* and *Zygophyllum coccineum,* as well as 6 minor communities of *Salicornia herbacea, Juncus rigidus, Phragmites australis, Aeluropus lagopoides, A. littoralis* and *Cressa cretica.*

The stands which represent the above community types vary in surface area from about 30 m² to tens of square kilometres. Often, they are dominated by a single species, but sometimes they form mixed communities in which 2 or more species codominate, e.g. *Halocnemum–Seidlitzia* community or *Seidlitzia–Zygophyllum–Traganum nudatum–Nitraria* community.

The vegetation of these marshes shows distinct zonation. Different communities occupy belts which run parallel to the shoreline. Zonation differs from one site to another and appears to be influenced by the horizontal distance from the shoreline and ground level in relation to high water mark. These factors determine the frequency and duration of inundation by sea water. The depth of the saline water table also affects zonation probably through influencing the moisture and salt content of the soil. For example, the *Halocnemum* zone generally occurs on the lower marshes. Its seaward edge is inundated by tides about 440 times per year, whereas the landward edge is covered only 10 times per year. Frequent inundation would reduce soil salinity to levels near that of the Arabian Gulf waters (4.1–4.2%). Salinity of the surface soil is about 10% and decreases rapidly to about 1.2% further down. The water table usually occurs at a depth of 60–80 cm below the surface. A blue-black subsurface horizon which is rich in sulphides is characteristic of the lower marsh.

The middle marsh lies beyond the reach of the highest spring tides but appears to be mainly under the influence of the saline water table. Since evaporation is much greater than precipitation, soil water moves upwards by capillarity and deposits salt as a white crust on the surface. Salinity varies from 10 to 20% at the surface, and drops sharply to 0.4–1.5% in the lower horizons.

On the upper marshes, the saline water table is too deep to influence soil properties to a great extent. Moreover, the soil is predominantly of coarse sand so that capillary forces are weak. Soil salinity is fairly low, being 1–2% at the surface and 0.1–0.4% below.

The width of the vegetation belts as well as the zonal sequence seem to be influenced mainly by variations in ground level. Where the slope is gentle, the zones are very wide and merge very gradually into one another but when the slope is steep, zones are quite narrow and sharply distinct from each other. Local differences in microtopography may modify the general zonal sequence, e.g. when a sandy ridge occurs on the lower marsh, it often shelters species of the middle marsh.

Figure 7 shows a profile transect of a representative salt marsh at Al-Talh in the southwestern corner of Kuwait Bay (Fig. 3). It reveals the zonal pattern as well as the main components of each community and their percentage cover.

Considering the coastal salt marshes as a whole, their general zonal sequence is as follows:

- The low frequently inundated mudbanks and the creeks which drain the lower marsh are dominated by pure stands of *Salicornia herbacea* (Photo 2). This species appears to be well adapted to water-logged conditions and frequent submersion.
- The lower marsh is usually covered by a pure community of *Halocnemum strobilaceum* (Photo 3), but sometimes there are also pure or mixed communities of *Juncus rigidus, Phragmites australis* (Photo 4), *Aeluropus lagopoides, A. littoralis, Cressa cretica* and *Seidlitzia rosmarinus.* Species diversity is very low and plants are halophytic perennials. Coverage varies from 5% or less to 90%.

  It should be noted that recently, as a result of dumping of sewage on parts of the lower marsh, profound changes have taken place. *Salicornia herbacea* is almost entirely wiped out, *Juncus rigidus* and *Halocnemum strobilaceum* are very adversely affected, whereas *Phragmites australis*

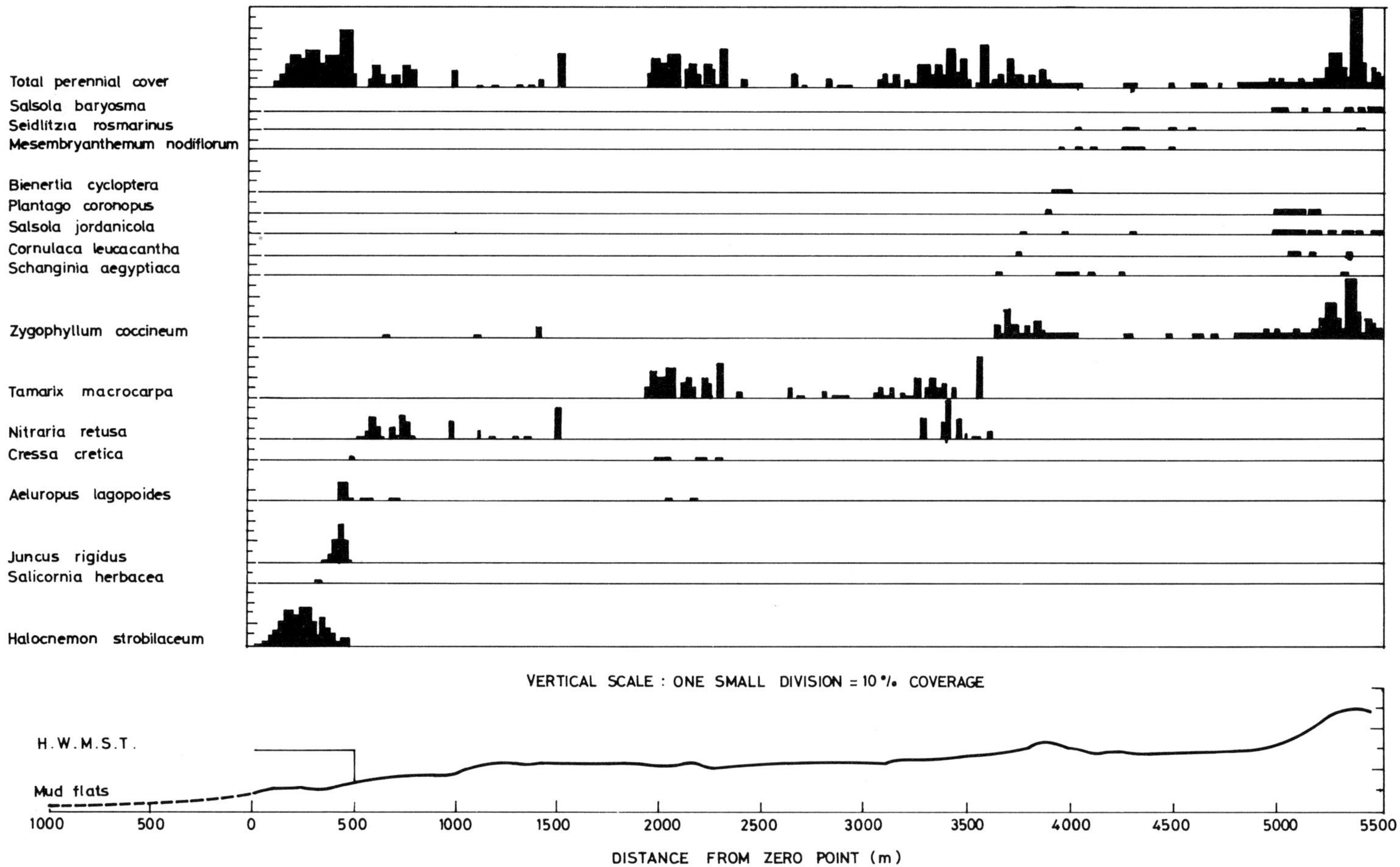

Fig. 7 Profile transect and plant coverage in a salt marsh at Al-Talh (HALWAGY/HALWAGY 1977).

has expanded its area tremendously and assumed gigantic dimensions, reaching a height of 4–5 m as compared to 1.2–1.5 m on sites not receiving sewage effluents.

- The middle marsh is often occupied by communities of *Nitraria retusa*, *Tamarix macrocarpa* (Photo 5), together with pure or mixed communities of *Seidlitzia rosmarinus*, *Aeluropus lagopoides* and *Traganum nudatum*. Associated species include *Suaeda vermiculata*, *Salsola baryosma*, *S. jordanicola*, *Schanginia aegyptiaca*, *Bienertia cycloptera*, *Frankenia pulverulenta* and *Mesembryanthemum nodiflorum*. All species are halophytic perennials or annuals and total coverage ranges from less than 5% to about 45%.
- The upper marsh is dominated by *Zygophyllum coccineum* (Photo 6). The soil is coarse sandy with low salinity. This improvement in soil conditions is reflected in greater species diversity, including many glycophytes, e.g. *Ifloga spicata*, *Plantago ovata*, *Filago desertorum* and *Schismus barbatus*. Total coverage ranges from 1% to 35%. On favoured sites, i.e. those receiving wind-drifted sand, sizeable phytogenic mounds may be formed on which coverage may reach 70%.

4.3.3 The desert plain and desert plateau ecosystems

As previously mentioned, the desert plain merges very gradually into the desert plateau. Both the plain and plateau shelter 4 communities: a relic community of *Panicum turgidum* and three communities of *Cyperus conglomeratus*, *Rhanterium epapposum* and *Hammada salicornica* (Fig. 6).

Community of *Panicum turgidum*

This small community occurs mainly around Al-Atraf and is subject to intensive grazing. *Panicum* is a highly palatable species and is continuously curtailed by the large sheep and goat population of the neighbouring village of Al-Jahra. It has often been suggested that the vegetation around village perimeters is the first to disappear. Similar observations were reported from the Sudan, where *Panicum* was almost completely wiped out from large areas around Omdurman as a result of overgrazing and intensive cutting by the local inhabitants for use as fuel (HALWAGY 1962).

Evidence from old vegetation maps (e.g. DICKSON 1955) shows that the community of *Panicum* has retreated considerably. In a quantitative study along a 99 km east-west transect extending from Al-Atraf to Al-Salmy (Fig. 3), EL-ABDULLAH (1979) estimated that *Panicum* had a frequency of 4%. The average total perennial cover was also 4%, indicating the extreme sparseness of the vegetation. Associated species include a few perennials, e.g. *Moltkiopsis ciliata*, *Stipagrostis plumosa* and *Asthenatherum forsskalii* and many tiny annuals, e.g. *Ifloga spicata*, *Filago desertorum*, *Crucianella membranacea*, *Loeflingia hispanica*, *Lophochloa pumila*, *Cutandia memphitica*, *Schismus barbatus*, *Lotus halophilus* and *Plantago ciliata*. A complete list of species of the *Panicum* community, together with their frequency (EL-ABDULLAH 1979) is given in table 2.

Species	% Frequency
Perennials	
Panicum turgidum FORSSK.	4
Rhanterium epapposum OLIV.	4
Astragalus spinosus (FORSSK.) MUSCHL.	1
Small perennials	
Moltkiopsis ciliata (FORSSK.) JOHNST.	87
Stipagrostis plumosa (L.) MUNRO EX T. ANDERS.	25
Asthenatherum forsskalii (VAHL) NEVSKI	18
Launaea mucronata (FORSSK.) MUSCHL.	4
Helianthemum lippii (L.) PERS.	1
Polycarpaea repens (FORSSK.) ASCHERS. ET SCHW.	1
Fagonia glutinosa DEL.	1
Annuals and biennials	
Ifloga spicata (FORSSK.) SCH.-BIP.	91
Filago desertorum POMEL	76
Crucianella membranacea BOISS.	73
Loeflingia hispanica L.	71
Lophochloa pumila (DESF.) BOR	70
Cutandia memphitica (SPRENG.) BENTH.	63
Schismus barbatus (L.) THELL.	59
Lotus halophilus BOISS. ET SPRUN.	50
Plantago ciliata DESF.	46
Silene arabica BOISS.	27
Astragalus schimperi BOISS.	25
Stipa capensis THUNB.	21
Neurada procumbens L.	20
Medicago laciniata (L.) MILL.	17
Launaea capitata (SPRENG.) DANDY	12
Ononis serrata FORSSK.	10
Picris babylonica HAND.-MAZZ.	10
Trachynia distachya (L.) LINK	9
Astragalus annularis FORSSK.	7
Paronychia arabica (L.) DEL.	5
Astragalus hauarensis BOISS.	4
Schimpera arabica HOCHST. ET STEUD.	4

Tab. 2 Floristic composition of community of *Panicum turgidum* around Al-Atraf (EL-ABDULLAH 1979).

Annuals and biennials/contd.

Species	% Frequency
Asphodelus viscidulus Boiss.	4
Plantago boissieri Hausskn. et Bornm.	4
Trigonella stellata Forssk.	4
Hippocrepis bicontorta Lois.	3
Plantago psammophila Agnew et Chalabi-Ka'bi	3
Erodium cicutarium (L.) L'Her.	3
Plantago ovata Forssk.	2
Herniaria hirsuta L.	2
Monsonia nivea (Decne.) J. Gay	1
Gastrocotyle hispida (Forssk.) Bge.	1
Anthemis deserti Boiss.	1
Lappula spinocarpos (Forssk.) Aschers.	0.2
Emex spinosus L.	0.2
Koelpinia linearis Pall.	0.2

The soil consists of fairly loose and easily drifted coarse sand (90–98% sand). The surface soils are mostly sands, rarely loamy sands, whereas the subsurface soils are usually sands, less frequently loamy sands and rarely sandy loams. Organic carbon content is very low (0.025–0.070% on the surface, 0.01–0.35% in the subsurface). Total dissolved solids (TDS) range from 0.2–1.7% and show no distinct pattern with depth. pH value ranges mostly from 7.2 to 7.8 and tends to increase with depth.

Community of *Cyperus conglomeratus*

This community was studied by Halwagy et al. (1982). It occupies a considerable area to the south and southwest of Kuwait City as well as a small pocket in the southeastern part of the country. The soil is loose coarse sand which is easily drifted by wind. Surface soils consist of about 90% sand, 7% silt and 3% clay. Organic carbon content is very low (0.075% ± 0.007%). The percentage of TDS is also low (0.05 ± 0.007). Soil reaction is alkaline (pH 7.44 ± 0.04).

Cyperus conglomeratus (Photo 7) is the dominant perennial. Inspite of its high frequency (90%) it has very low coverage, often between 1 and 5%, rarely reaching 15%. It is associated with a few smaller perennials, e.g. *Moltkiopsis ciliata*, *Launaea mucronata*, *Stipagrostis plumosa* and *Polycarpaea repens*. The community also shelters many spring annuals, e.g. *Lotus halophilus*, *Ifloga spicata*, *Lophochloa pumila*, *Schismus barbatus*, *Cutandia memphitica* and *Plantago ciliata*. *Launaea capitata*, a biennial species, is also common. The floristic composition of this community, as well as that of the *Rhanterium epapposum* and the *Hammada salicornica* communities are shown in table 3 (Halwagy et al. 1982).

Species	*Cyperus* community % Frequency	*Rhanterium* community % Frequency	*Hammada* community % Frequency
Perennials			
Cyperus conglomeratus Rottb.	90	–	–
Rhanterium epapposum Oliv.	–	50	–
Hammada salicornica (Moq.) Iljin	–	–	43
Calligonum comosum L'Her.	–	0.5	–
Zilla spinosa (L.) Prantl	–	–	4
Astragalus spinosus (Forssk.) Muschl.	–	–	4
Cornulaca leucacantha Charif et Aellen	–	–	2
Convolvulus oxyphyllus Boiss.	–	–	0.4
Small perennials			
Moltkiopsis ciliata (Forssk.) Johnst.	38	13	–
Launaea mucronata (Forssk.) Muschl.	18	14	10
Stipagrostis plumosa (L.) Munro ex T. Anders.	15	40	20
Polycarpaea repens (Forssk.) Aschers. et Schw.	5	2	0.4
Fagonia sp.	–	3	–
Fagonia brugieri DC.	–	–	1
Herniaria hemistemon J. Gay	–	–	9
Annuals and biennials			
Lotus halophilus Boiss. et Sprun.	89	55	–
Ifloga spicata (Forssk.) Sch.-Bip.	82	24	48
Lophochloa pumila (Desf.) Bor	80	37	56
Schismus barbatus (L.) Thell.	70	84	76
Cutandia memphitica (Spreng.) Benth.	69	89	1
Plantago ciliata Desf.	63	15	5
Launaea capitata (Spreng.) Dandy	61	48	9
Loeflingia hispanica L.	43	–	0.4
Filago desertorum Pomel	43	5	15
Plantago boissieri Hausskn. et Bornm.	40	23	0.4
Astragalus schimperi Boiss.	39	15	13
Crucianella membranacea Boiss.	31	2	1
Trachynia distachya (L.) Link	30	–	–
Medicago laciniata (L.) Mill.	25	7	12
Silene arabica Boiss.	20	8	6
Asphodelus viscidulus Boiss.	18	9	–
Stipa capensis Thunb.	15	27	20
Astragalus annularis Forssk.	12	–	7
Picris babylonia Hand.-Mazz..	8	23	20
Koelpinia linearis Pall.	7	11	4
Bupleurum semicompositum L.	7	–	–

Tab. 3 Floristic composition of communities of *Cyperus conglomeratus*, *Rhanterium epapposum* and *Hammada salicornica* in the desert plain and desert plateau (Halwagy et al. 1982).

Annuals and biennials/contd.

Species	*Cyperus* community % Frequency	*Rhanterium* community % Frequency	*Hammada* community % Frequency
Astragalus hauarensis Boiss.	5	13	0.4
Paronychia arabica (L.) Del.	5	1	6
Erodium laciniatum (Cav.) Willd.	3	0.5	3
Ononis serrata Forssk.	3	13	–
Brassica tournefortii Gouan	2	–	–
Plantago ovata Forssk.	2	14	2
Atractylis cardus (Forssk.) Christ.	2	4	1
Hippocrepis bicontorta Lois.	0.4	3	1
Silene villosa Forssk.	0.4	32	0.4
Lappula spinocarpos (Forssk.) Aschers. et Kuntze	0.4	–	6
Helianthemum ledifolium (L.) Mill.	–	17	1
Malcolmia grandiflora (Bge.) O. Ktze.	–	10	3
Arnebia decumbens (Vent.) Cross. et Kral.	–	7	30
Neurada procumbens L.	–	7	3
Haplophyllum tuberculatum (Forssk.) Adr.-Juss.	–	6	–
Anthemis deserti Boiss.	–	5	2
Schimpera arabica Hochst. et Steud.	–	3	38
Asphodelus fistulosus L. var. *tenuifolius* Cav.	–	3	0.4
Gymnarrhena micrantha Desf.	–	1	2
Plantago psammophila Agnew et Chalabi-Ka'bi	–	1	1
Emex spinosus L.	–	0.5	1
Gypsophila antari Post	–	0.5	1
Trigonella stellata Forssk.	–	0.5	8
Carduus pycnocephalus L.	–	0.5	–
Gastrocotyle hispida (Forssk.) Bge.	–	0.5	8
Spergularia diandra (Guss.) Heldr. et Sart.	–	–	12
Oligomeris linifolia (Vahl) Macbride	–	–	8
Senecio desfontainei Druce	–	–	7
Arnebia tetrastigma Forssk.	–	–	4
Calendula aegyptiaca Desf.	–	–	4
Savignya parviflora (Del.) Webb	–	–	3
Hordeum geniculatum All.	–	–	3
Reseda arabica Boiss.	–	–	3
Reichardia tingitana (L.) Roth.	–	–	2
Hypecoum pendulum L.	–	–	1
Torularia torulosa (L.) O.E. Schulz	–	–	1
Atriplex dimorphostegia Kar. et Kir.	–	–	1
Rumex pictus Forssk.	–	–	0.4
Bassia eriophora (Schrad.) Aschers.	–	–	0.4
Melilotus parviflora Desf.		–	0.4

Community of *Rhanterium epapposum*

This community covers considerable areas in the centre, south and extreme northeast of the country (Fig. 6). According to HALWAGY et al. (1982), soils are rather firm coarse sand (87%) with a sizeable proportion of silt (about 11%) and little clay (2%). In common with soils of arid regions, these soils are poor in organic carbon (0.05% ± 0.007%), and slightly alkaline (pH 7.59 ± 0.05). They have higher content of TDS (0.22% ± 0.004%) as compared to those of the *Cyperus conglomeratus* community. *Rhanterium epapposum* (Photo 8), the dominant perennial, has a frequency of 50% and a coverage of 1-25%, rarely reaching up to 50%. It seems to flourish most when it receives a reasonable amount of drifted sand. It is associated with a number of subordinate perennials, e.g. *Stipagrostis plumosa*, *Launaea mucronata* and *Moltkiopsis ciliata*. Numerous small annuals and a few biennials occupy the ground layer, e.g. *Cutandia memphitica*, *Schismus barbatus*, *Lotus halophilus* and *Launaea capitata*.

Community of *Hammada salicornia*

This is by far the largest community in surface area and occurs in the west and north of the country. Soils are usually thin, with the subsurface hardpan frequently outcropping on the surface. Soil is generally firm and mechanical analysis shows that it consists of about 90% sand, 6% silt and 4% clay. Soil reaction is mildly alkaline (pH 7.48 ± 0.05). It is also poor in organic carbon (0.047% ± 0.005%) but contains appreciable quantities of TDS (0.35% ± 0.05%).

Hammada salicornica (Photo 9) has a frequency of 43% and a coverage of 1–25%. On exceptionally favourable sites, i.e. those receiving considerable quantities of blown sand, *Hammada* can build sizeable mounds and coverage may reach up to 33%. Subordinate perennials include *Stipagrostis plumosa* and *Launaea mucronata*. The community also supports many annuals, e.g. *Schismus barbatus*, *Lophochloa pumila* and *Ifloga spicata*.

The community exhibits considerable heterogeneity. Whereas *Hammada* becomes particularly luxuriant on deep deposits of loose drifted sand, it may disappear entirely where the hardpan is exposed. On such sites, the ground is either completely barren or occupied by almost pure stands of *Stipa capensis* which look very much like cultivated fields of wheat (Photo 10). Occasionally, a thin veneer (c. 10 cm) of sand overlies the hardpan and the vegetation is then made almost entirely of short annuals, e.g. *Schismus barbatus*, *Arnebia decumbens* and *Lophochloa pumila*.

5. Conclusions

From the foregoing pages, it is clear that the landscape of Kuwait is treeless (except for cultivated trees). The natural vegetation includes a few bushy short shrubs less than 2 m in height (*Lycium shawii*, *Nitraria retusa* and *Tamarix macrocarpa*). The main life form consists of shorter plants (often below 1 m high). These include hardy or succulent undershrubs and perennial herbs, e.g. *Rhanterium epapposum*, *Hammada salicornica*, *Zygophyllum coccineum*, *Seidlitzia rosmarinus*, *Atriplex leucoclada*, *Aellenia glauca*, *Suaeda vermiculata.*, *Convolvulus* spp., *Astragalus* spp., *Salsola baryosma*, and *Cornulaca leucacantha*, as well as perennial grasses, sedges and rushes (*Panicum turgidum*, *Pennisetum divisum*, *Cyperus conglomeratus* and *Juncus rigidus*). The understory is occupied by a few short perennials as well as many small spring ephemerals.

The major habitat types are the coastal sand dunes, coastal salt marshes, desert plain and desert plateau. They reveal sharp differences in soil conditions and support distinct vegetation types. Tallest vegetation and highest coverage usually occur on the salt marshes while the desert plain and plateau support the thinnest cover (of shorter plants). Each habitat type is characterized by a certain assemblage of dominant and associated species. Within each habitat type there are several minor habitat subtypes which reveal obvious or subtle differences between them. When these differences are great as for example on the salt marshes or sand dunes, the corresponding communities which these habitat subtypes support are also widely different. They usually differ in species diversity, the name of the dominant and to some extent also the associated species, total coverage and plant height. In general, species diversity is very low when the habitat conditions are extreme since only pioneer well adapted species can live under such difficult conditions. On the other hand, when the differences between the habitat subtypes are small, e.g. on the desert plain and plateau, the corresponding communities reveal only minor differences. These are usually limited to differences in community spectrum and change of dominance, with less obvious differences in total coverage and practically no difference in stature.

6. References

DICKSON, V., The wild flowers of Kuwait and Bahrain, 1955, London.

EL-ABDULLAH, R. H., A study of plant ecology along a transect from Al-Jahra to Salemy, Kuwait, M. Sc. dissertation, University of South Carolina, 1979.

ERGUN, H., Reconnaissance soil survey, Report to the Government of Kuwait, FAO/KU/TF17, 1969, Rome.

FUCHS, W. / GATTINGER, E. T. / HOLZER, F. H., Explanatory text to the synoptic geological map of Kuwait, Geological survey of Austria, 1968, Vienna.

Halwagy, M., Ecological studies of the desert vegetation of Kuwait with especial reference to the salt marshes, M. Sc. dissertation, University of Kuwait, 1973.

Halwagy, R., On the ecology of *Psamma arenaria* and *Carex arenaria* on the sand dunes at Blakeney Point, Norfolk, Ph. D. dissertation, University of London, 1953.

Halwagy, R., The incidence of the biotic factor in northern Sudan, Oikos *13*, 1962, 98–117.

Halwagy, R. / Halwagy, M., Ecological studies on the desert of Kuwait, I. The physical environment, J. Univ. Kuwait (Sci.) *1*, 1974 a, 75–86.

Halwagy, R. / Halwagy, M., Ecological studies on the desert of Kuwait, II. The vegetation, J. Univ. Kuwait (Sci.) *1*, 1974 b, 87–95.

Halwagy, R. / Halwagy, M., Ecological studies on the desert of Kuwait, III. The vegetation of the coastal salt marshes, J. Univ. Kuwait (Sci.) *4*, 1977, 33–74.

Halwagy, R. / Moustafa, A. F. / Kamel, S. M., On the ecology of the desert vegetation in Kuwait, J. Arid Environments, *5*, 1982, 95–107.

Kernick, M. D., Pasture, fodder production and range management project for sheep production in the region of Jahra, Kuwait, Ministry of Public Works, 1964, Kuwait.

Kernick, M. D., Plant resources, range ecology and fodder plant introduction, FAO, TA 2188, 1966, Rome.

Macksad, A., The desert flowers of Kuwait, Al-Arabi *132*, 1969, 94–103, Ministry of Information, Kuwait.

Milton, D. I., Geology of the Arabian Peninsula, Kuwait, U.S. Geol. Survey Prof. Paper 560-F, 1967, 1–7.

Omar, S. A. / Taha, F. K. / Nassef, A. A. / al-Fili, K., Animal counts of Kuwait's rangelands, KISR Annual Research Report, 1981, 20–22, Kuwait Institute for Scientific Research, Kuwait.

Picha, F. / Saleh, A. A. M., Quaternary sediments in Kuwait, J. Univ. Kuwait (Sci.) *4*, 1977, 169–185.

Tansley, A. G., The British Islands and their vegetation, Vol. 2, 1965, Cambridge University Press.

UNESCO, Map of the world distribution of arid regions, MAB Technical Notes *7*, 1979, UNESCO, Paris.

USDA, Soil Survey Staff, Soil classification, A comprehensive system, Seventh approximation, USDA, 1960, Washington.

Walter, H., Vegetation of the earth, English edition, 1973, New York.

Walter, H. / Harnickell, E. / Mueller-Dombois, D., Klimadiagramm-Karten der einzelnen Kontinente und die ökologische Klimagliederung der Erde, 1975, Stuttgart.

Acknowledgements

I wish to express my sincere thanks to Dr. Azza Al-Musallam, Head, Department of Botany and Microbiology for providing various facilities. I am also grateful to Mr. M. B. Abdulrazzaq and Mr. M. K. Azhar for typing the manuscript, Mr. A. R. Jassim for drawing the figures and Mr. S. Rajendra in connection with photographic work. Mr. A. M. Al-Kulaib, Superintendent of Climatology, and his staff kindly supplied the meteorological data.

Prof. Dr. Riad Halwagy, Department of Botany and Microbiology, Faculty of Science, University of Kuwait, P. O. Box 5969, Kuwait.

Photo 1 *Pennisetum divisum* on the coastal oolitic dunes at Al-Dba'iyyah.

Photo 2 Creek draining the lower salt marsh at Al-Talh. Note *Salicornia herbacea* on sloping mudbank in the foreground, *Halocnemum strobilaceum* on higher ground (background).

Photo 3 Community of *Halocnemum strobilacaeum* on the lower salt marsh at Al-Talh.

Photo 4 Community of *Juncus rigidus* (short) and *Phragmites australis* (tall) on the lower marsh at Al-Talh.

Photo 5 Communities of *Tamarix macrocarpa* (foreground) and *Nitraria retusa* (background) at Al-Talh.

Photo 6 Community of *Zygophyllum coccineum* at Al-Talh. *Zygophyllum* flourishes on blown sand and builds sizeable mounds. In the background are plantations of *Tamarix* spp.

Photo 7 Community of *Cyperus conglomeratus* on loose coarse sand. Note the numerous short annuals.

Photo 8 Community of *Rhanterium epapposum*. Dense annuals occur on more fixed soil (background).

Photo 9 Community of *Hammada salicornica. Hammada* is short-grazed. Numerous annuals can be seen.

Photo 10 *Stipa capensis* forms almost pure stands where soil is thin and gritty.

Heinz Ullrich Baierle und Wolfgang Frey

A vegetational transect through Central Saudi Arabia (aṭ-Ṭā'if – ar-Riyāḍ)

Contents

1. Introduction

The extensive desert and semi-desert landscapes of central Saudi Arabia are the nucleus of the eastern Saharo-Arabian subregion of the Saharo-Arabian floristic region. This central part of Saudi Arabia deserves particular attention in vegetation ecology because most of the vegetation units of this phytogeographical region are represented here. In addition, there are two distinct geological provinces, the Arabian Shelf and the Arabian Shield, and this division is clearly reflected in the species composition of the vegetation units. This area will also surely be important in the future with regard to natural resources: there are a variety of geological substrata, on which the development and regeneration of numerous different vegetation units can be studied.

Thus far there have been a great many gaps in geobotanical research on central Saudi Arabia. Although the studies by MIGAHID (1978) and DE MARCO/DINELLI (1974) have provided relatively good information about the species that occur here, for most of the area there is a lack of detailed investigations on the characteristic vegetation units and their distribution. Some studies, such as those by EL-WAILLY (1972) and MANDAVILLE (1965), are concerned with very special subjects or limited areas, while others, such as that by MIGAHID/EL-SHEIKH (1977), give only a very general survey of vegetation conditions.

Detailed studies of the vegetation in the extremely arid regions of Saudi Arabia have been undertaken only in the al-Aḥsā' province (GIACOMINI et al. 1979 a, b) and in small parts of al-Ḥiǧāz (BATANOUNY 1979, BATANOUNY/BAESHIN 1983). Even the studies by VESEY-FITZGERALD (1957 a, b), which provide good comprehensive surveys, are basically descriptive in nature. These publications, as well as the others named above, lack tables showing the exact make-up of the vegetation units described.

The present study describes the vegetation units along a transect from aṭ-Ṭā'if to ar-Riyāḍ, thus contributing to a better understanding of the vegetation ecology of Saudi Arabia's central desert areas. This is a condensed English version of a comprehensive description of this transect in BAIERLE/EL-SHEIKH/FREY (1985). An overall picture of the vegetation structure, including the areas between aṭ-Ṭā'if and Ǧidda and between ar-Riyāḍ and Dharān, is given in the map published by BAIERLE et al. (1985).

The field work was carried out in March, 1982, as part of a research project undertaken jointly by the botanical working group of the Special Research Division 19 TAVO, the Saudi Biological Society (SBS) and the Saudi Arabian National Center for Science and Technology (SANCST). We are grateful to these organizations for their financial support, to Prof. Dr. W. Röllig (TAVO) for his continuous assistance, and to Dr. A. M. El-Sheikh, Prof. Dr. A. M. Migahid (both of the Botany Department, King Saud University, Riyadh) and Dipl.-Biol. P. König (Berlin), who accompanied us in our travels. For identification of the plant material we are indebted to Mr. P. König, Dr. H. Kürschner and Prof. Dr. H. Scholz (all from Berlin) and to Mr. I. C. Hedge and the other specialists of the Royal Botanic Garden, Edinburgh. We also thank Mr. M. Seidel for the translation.

2. Location, geological setting and climatic conditions in the study area

Location

The study area is situated in the western half of the central part of the Arabian Peninsula. It comprises a strip up to 100 km in width along the road from aṭ-Ṭā'if to ar-Riyāḍ (northern route): aṭ-Ṭa'if (40° 25′ E,

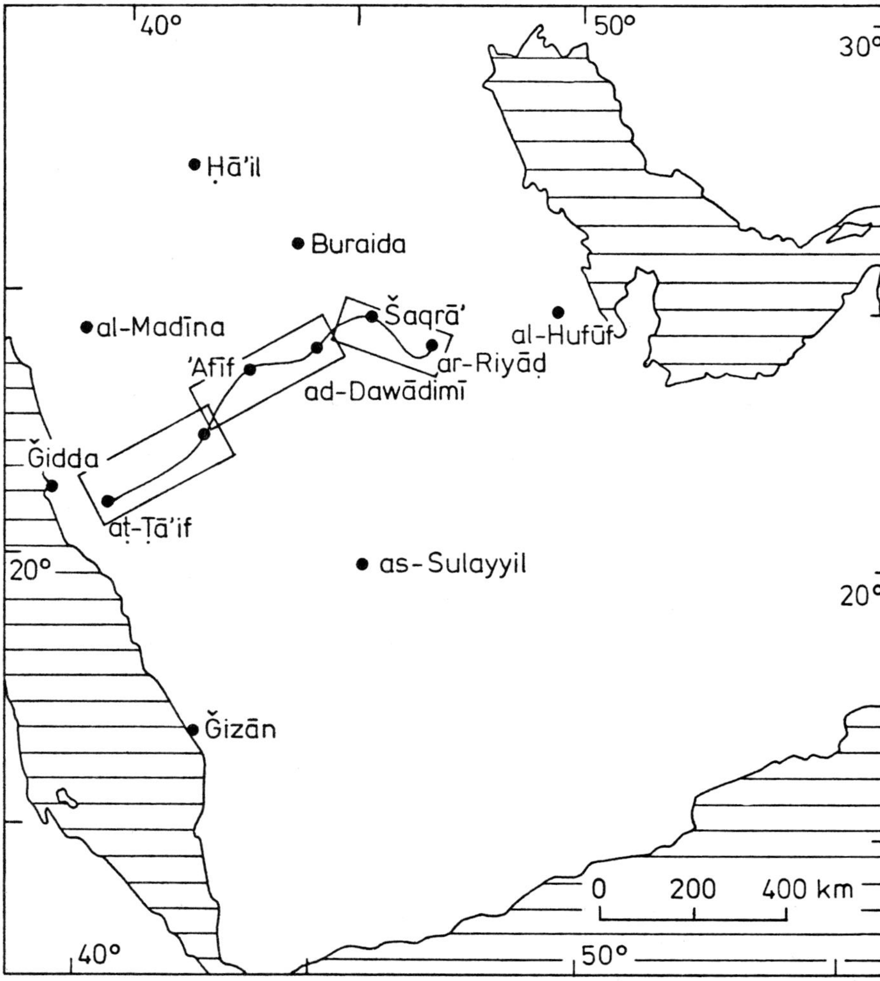

Map 1 Location of the investigated area

21° 16′ N) – Ẓalim (42° 11′ E, 22° 43′ N) – 'Afīf (42° 56′ E, 23° 55′ N) – ad-Dawādimī (44° 23′ E, 24° 30′ N) – Šaqrā' (45° 15′ E, 25° 16′ N) – ar-Riyāḍ (46° 43′ E, 24° 40′ N).

Geological conditions and natural units

The road from aṭ-Ṭā'if to ar-Riyāḍ traverses the Arabian Shield and the Arabian Shelf, the structural provinces which together constitute the geologically stable land mass of the peninsula. The western part of this stable land mass is formed by the Precambrian basement, which consists of crystalline and highly metamorphosed rocks and is part of the ancient Afro-

Asian craton (Arabian Shield). In the shelf area adjacent to the east this Precambrian plate is covered by sediment layers (limestones, sandstones, marls, etc.) that are thick in places and are predominantly of Mesozoic and Neozoic origin (on geology see AL-SAYARI/ZÖTL [1978] and POWERS et al. [1966]).

On the basis of these major geological divisions and the surface-forming processes of the past a number of natural units can be distinguished in central Saudi Arabia. They are in a roughly N-S arrangement (cf. map in BARTH 1976; Fig. 8). The W-E transect crosses al-Ḥiǧāz, Naǧd pediplain, and the Arabian cuesta region.

The western edge of the study area is formed by the eastern slopes of al-Ḥiǧāz, which is the middle section of the western border range, running from NNW to SSE. These mountains, which have a very irregular relief, are made up mainly of granitic gneisses, crystalline slates and gneissic granodiorites, with a small portion of massive coarse-crystalline granites.

The Naǧd pediplain meets the base of al-Ḥiǧāz at about 1500 m above sea level and dips to the east. In contrast to the western border range it has only a moderate relief. An extensive peneplain landscape (made up of crystalline slates, gneiss, granite-gneiss, chlorite schists, andesite, rhyolite and granites) has developed here as a result of erosion processes that have been active since the Precambrian (coalescing pediments and desert plains interrupted here and there by inselbergs [BROWN 1960]). Extensive lava flows (harrahs) in the western highlands are a result of volcanic activity.

The Arabian cuesta region begins east of the crystalline zone (in the study area east of ad-Dawādimī). Here Palaeozoic, Triassic, Jurassic and more recent strata meet the surface, extending in broad belts from NNW to SSE. Some of these strata form cuestas of various heights (such as the Tuwayq escarpment). The scenery created by these exposed bedding planes is that of a rock desert. Other, usually lower-lying bedding planes are partially covered by onblown sand or are overlain by a closed sand cover. Such sands and other Pleistocene accumulations of alluvial origin characterize large parts of the landscape, especially in the subsequent lowlands along the scarps.

Between some of the cuestas there are elongated fields of sand dunes, which constitute a separate type of landscape. In the study area Nafūd Qunaifiḏa and Nafūd as-Sirr are such regions. Here large hills of sand predominate, along with various forms of long, narrow and usually parallel sand ridges and dune chains (uruqs).

Drainage is generally towards the east. On account of the very low and irregular precipitation and the high rates of evaporation there is a complete lack of perennial drainage systems. The broad valley systems, such as Wādī r-Rišā', probably developed during periods with more precipitation in the Pleistocene. Some areas are not connected with valley systems and drain into depressions without outlets. In their terminal basins there are often saline flats (sebkhas).

Soils

Owing to the high aridity and the poverty of vegetation in the central desert areas there are no of the well-developed profiles such as found in many soils in more humid climates. Typically are here soils in the beginning stage of development, having been subjected mainly by physical weathering processes. BARTH (1976) divides these lithomorphic soils into three groups, depending on the parent material: Lithosols, Regosols and Dust Yerma. In addition, there are Solonchaks and Solonezes in inland basins without outlets. Finally, in oasis areas Vertisols with very clayey surface layers can occur.

Climatic conditions

The central areas of Saudi Arabia are characterized by an arid to extremely arid climate. The mean annual precipitation in the study area is 100 to 150 mm east of al-Ḥiǧāz (meteorological stations at ar-Riyāḍ, ad-Dawādimī and ʿAfīf) and 200 to 240 mm near the eastern al-Ḥiǧāz slopes. In the central desert areas precipitation occurs only in the winter and spring months, and in the southern al-Ḥiǧāz region it occurs mainly during this period (maximum precipitation in March and April). Rain is usually in the form of heavy showers limited to certain localities. As in other arid regions the annual precipitation varies considerably from year to year.
The differences between the mean temperatures of the coldest and the warmest months are much smaller in the al-Ḥiǧāz region than in the interior with its hot summers and relatively cool winters. The same is true as a rule for the differences between temperature extremes.

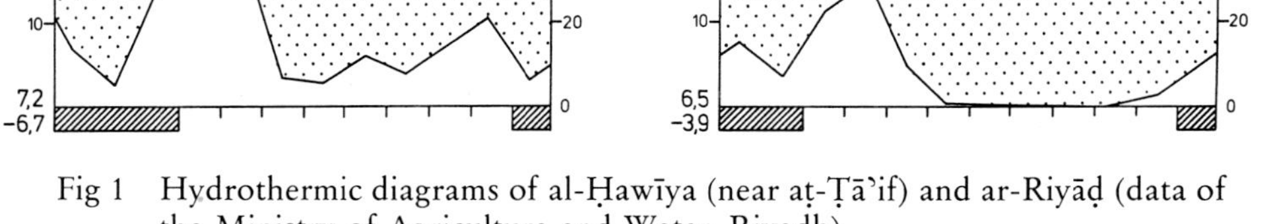

Fig 1 Hydrothermic diagrams of al-Ḥawīya (near aṭ-Ṭāʾif) and ar-Riyāḍ (data of the Ministry of Agriculture and Water, Riyadh).

3. Methods

In the vegetation units of the study area a total of 115 vegetation samples were carried out using the method of Braun-Blanquet. The samples were tabulated according to the formation classes (Frey/Probst 1977) to which the units belonged.

Demarcation of the vegetation units and naming of the communities was generally based on the characteristic perennial species, following the method of Zohary (1973, 1982) and other investigators of plant communities in southwest Asia. The term "community" should be understood here in an unranked sense. Character species of a community can, but need not also be character species of an association as defined by Braun-Blanquet. This departure from Braun-Blanquet's system of evaluation is necessary in view of the present state of phytosociological research on the Arabian Peninsula.

Three tables are reproduced in this study. The complete tabular material and a detailed description of the methods can be found in Baierle/El-Sheikh/Frey (1985).

4. Vegetation

4.1 Dwarf-scrub and related communities, grasslands

In central Saudi Arabia most of the areas whose physiognomy is characterized by vegetation are occupied by dwarf-scrub and related communities. This formation class is typical of sandy plains, stony and gravelly surfaces with patches of sand, and, to a lesser extent, purely stony habitats with a good water supply such as scarp slopes in the cuesta region. The coverage varies widely depending on substrate and relief. However, dwarf-scrub seldom covers more than 5–10% of the surface.

Among the differentiating factors for the distribution and form of the dwarf-scrub communities and grasslands are geographic location (and thus climatic conditions), the geology of the substrate, the form of the terrain, the soil profile (e.g. thickness of loose sand layers), and physical and chemical soil characteristics. In this respect the border between the shelf area and the crystalline basement is especially important. For this reason the extensive dwarf-scrub communities of the Arabian Shelf and the Arabian Shield will be discussed separately.

Grasslands cover only minor parts of the study area. Since the *Panicum turgidum* grasslands overlap with the dwarf shrublands in the western region, it seemed reasonable to deal with the communities of these two formation classes together.

4.1.1 Dwarf-scrub of the Arabian Shelf (Table 1)

The *Rhanterium epapposum* dwarf-shrublands (Photograph 1) are the most important vegetation unit of the cuesta region. They colonize several types of sites, which differ considerably with respect to topography and surface conditions. Stands occur in narrow strips along the margins of sterile rock deserts, in islandlike patches in vegetation-poor gravel and rubble plains, and over large areas extending over many square kilometres.

Rhanterium epapposum dwarf-shrublands cover areas where Quaternary gravel (Type 1, Samples 15–18) or Mesozoic consolidated sediments such as limestones and sandstones (Type 2, Samples 9–14) are covered with an unbroken layer of loose sand at least 30 cm thick. They are also found on many regs with patches of onblown sand (Type 3, Samples 5–7) and on some craggy, weathering-resistant cuesta scarps with northern exposures (Type 4, Samples 3 and 4).

The other dwarf-scrub communities occurring in the cuesta region are much less important than the *Rhanterium epapposum* dwarf-shrublands. Examples are the *Farsetia aegyptiaca* community and a community dominated by *Ephedra* spec. which together with the *Rhanterium epapposum* community (Type 4) make up the vegetation of steep faces (e.g. on wadi banks or scarp edges). The dominant species mentioned are accompanied on these steep rocky slopes by *Gymnocarpos decandrum*, *Stipagrostis plumosa*, *Fagonia bruguieri* and other species. Of these, *Gymnocarpos decandrum* is frequently co-dominant. This species also characterizes stands of the *Rhanterium epapposum* community that grow on a rocky substrate (Types 3 and 4), whereas it is absent in the *Rhanterium epapposum* community on closed sand covers (Types 1 and 2).

On the gypseous bedrock of the Jilh formation, which is covered with onblown sand in some places, *Rhanterium epapposum* is replaced by *Salsola cyclophylla* (Sample 8).

Where the hardened masses of pebble and gravel (symbol Qg_2 on the geological map) are covered with only a thin layer of sand (about 10 cm) there are dwarf-shrublands characterized by *Farsetia burtonae* (Samples 19–21).

Hammada salicornica dwarf-shrublands (Samples 22–24) colonize deep, level accumulations of loose Quaternary sand, which cover large areas in rift zones and in old or recent valley systems. The habitats of this community are similar in topography to many of the areas occupied by the *Rhanterium epapposum* dwarf-shrublands (Types 1 and 2). Soil analyses have shown, however, that *Hammada salicornica* habitats have noticeably higher salt contents than comparable *Rhanterium epapposum* habitats. Further studies will be necessary to determine whether there is a causal relation between the moderate salinity of the upper soil layers under *Hammada sal-*

Table 1 Dwarf-scrub of the Arabian shelf (without wadi vegetation).
Extract of table 3 in BAIERLE/EL-SHEIKH/FREY (1985).

Serial number:	1	2	3	4	5	6	7	8	9	10	11	12	13	14	15	16	17	18	19	20	21	22	23	24
Number of species:	16	4	6	11	22	10	15	16	13	9	6	5	9	5	12	17	19	13	15	12	9	8	6	11
Character species:																								
Farsetia aegyptiaca	1			+																				
Ephedra spec.		1	r																					
Rhanterium epapposum		(+)	1	1	1	1	1	+	1	2a	1	1	1	1	1	1	1	1						
Panicum turgidum						r	1																	
Lasiurus scindicus						+	1																	
Salsola cyclophylla								1																
Farsetia burtonae																			1	1	+			
Hammada salicornica																						1	2a	1
Character species of higher rank/differential species:																								
Gymnocarpos decandrum	1	1	1	1	+	r	+																	
Stipagrostis plumosa	1	+	1	+	1		1			+					+		1		1	+				
Fagonia bruguieri	2m	(+)	+	1												+		+	+					
Helianthemum lippii s.l.	1			r												1	+	+	1	2m	1			
Astragalus spinosus			r													+	+	+			+			
Heliotropium ramosissimum					r				r							+			+	1	+			

	Serial numbers 1–4	5–8	9–14	15–18	19–21	22–24
Geology of substratum:	Mesozoic sediments					
				Large-scale quarternary deposits		
Formation according to geological map:	Jubailah- u.Tuwayq-Form.	Marrat-,Minjur- und Jilh-Formation		Qg_2-Formation		Qu-Formation
Topography/Relief:	Steep slope (20°-30°)	Plain to moderate inclined slope (max. 15°)	Even or slight undulation			
Surface:	Rocky	Stony with fragmentary patches of sand	Compact layer of sand			
			About 40 cm deep		About 10 cm deep	Deeper than 1 m
Special characteristics:		Gypsum (serial no. 8)				Moderate surface saline crystallization

Locality of samples (to table 1)
(Locality, sea level, geological formation, slope inclination and exposition)

Sample		Locality	
1	(28):	5 km NE Ṯādiq;	720 m above sea-level, Tuwayq-formation; 25° SW
2	(4):	18 km SW ar-Riyāḍ;	725 m above sea-level, Jubailah-formation, 25° N
3	(3):	18 km SW ar-Riyāḍ;	725 m above sea-level, Jubailah-formation; 25° N
4	(27):	5 km NE Ṯādiq;	725 m above sea-level, Tuwayq-formation; 30° N
5	(51):	35 km WSW Šaqrā';	860 m above sea-level, Jilh-formation
6	(31):	7 km SW Šaqrā';	850 m above sea-level, Minjur-formation; 10° SSW
7	(30):	7 km SW Šaqrā';	850 m above sea-level, Marah-formation
8	(48):	28 km SW Šaqrā';	850 m above sea-level, Jilh-formation
9	(49):	20 km SW Šaqrā';	820 m above sea-level, Minjur-formation
10	(33):	9 km SW Šaqrā';	840 m above sea-level, Minjur-formation
11	(32):	9 km SW Šaqrā';	840 m above sea-level, Minjur-formation
12	(34):	9 km SW Šaqrā';	840 m above sea-level, Minjur-formation
13	(46):	31 km NE Ḫuff;	820 m above sea-level, Jilh-formation
14	(45):	31 km NE Ḫuff;	820 m above sea-level, Jilh-formation
15	(38):	18 km ESE Ḫuff;	810 m above sea-level, Qg_2-formation
16	(54):	15 km SSE Ḫuff;	850 m above sea-level, Qg_2-formation
17	(57):	38 km NE ad-Dawādimī;	840 m above sea-level, Qg_2-formation
18	(58):	38 km NE ad-Dawādimī;	840 m above sea-level, Qg_2-formation
19	(35):	18 km SE Ḫuff;	820 m above sea-level, Qg_2-formation
20	(36):	18 km SE Ḫuff;	820 m above sea-level, Qg_2-formation
21	(37):	18 km SE Ḫuff;	820 m above sea-level, Qg_2-formation
22	(9):	34 km SE ar-Riyāḍ;	600 m above sea-level, Qu-formation
23	(10):	34 km SW ar-Riyāḍ;	600 m above sea-level, Qu-formation
24	(11):	34 km SW ar-Riyāḍ;	600 m above sea-level, Qu-formation

icornica and the presence of this species or the absence of *Rhanterium epapposum.* It is conceivable that *Hammada salicornica*, with its long root system, is able to reach deeper, saline groundwater or moisture layers and that the measured soil salinity comes from shoots that have dropped off and decayed.

At the western border of the Arabian Shelf the extensive *Rhanterium epapposum* dwarf-shrublands reach their western limit of distribution (on wadi dwarf-shrublands of *Rhanterium epapposum* see the next section). This is confirmed by observations by Batanouny (cited in Thalen 1979) and Vesey-Fitzgerald (1957b). The geology of the subsoil presumably is of decisive importance for the distribution of *Rhanterium epapposum* dwarf-shrublands.

4.1.2 Dwarf-scrub and grasslands of the Arabian Shield (Table 2)

In contrast to the cuesta region, where *Hammada salicornica* dwarf-shrublands are of minor importance, the eastern Arabian Shield is characterized by open to very open xeromorphic dwarf-shrublands of this species (Photograph 2).

The total coverage of this community varies widely depending on relief, as does the size of the individual areas covered. In the broad valley plains (e.g. Wādī r-Rišā') and in the beds of the undrained depressions it forms large, unbroken stands covering many square kilometres. In areas with gentle or moderate relief *Hammada salicornica* dwarf-shrublands are usually confined to low-lying patches of sand and fine gravel in the middle of plantless rock or gravel deserts.
The subsoil consists of sands of various grain sizes, with high contents of gravel and fine gravel in some places. In many cases the soil is fairly compact. Under the surface layer of sand, at a depth of some 40 to 50 cm, very compact layers of loam or gravel can occur. The *Hammada* plants are situated on small hummocks, where they accumulate sand.
Moderate salinity is a common feature of soils colonized by *Hammada salicornica.* At least one horizon in the upper 50 cm has a sodium content of 50 ppm or more. The levels of other ions, particularly K, are also slightly higher.
Hammada dwarf-shrublands begin to become less common west and southwest of a line crossing the road from aṭ-Ṭā'if to ar-Riyāḍ about 20 km northeast of Ẓalim. Still further west this community is found only at occasional sites, while dwarf-shrublands and grasslands characterized by other species predominate.
West of Ẓalim *Salsola cyclophylla* dwarf-shrublands cover larger areas than in the cuesta region. Here, however, they grow mainly on sand and other fine sediments. In some of the large valley plains northeast of Ḥarrat Ḥaḍan *Salsola cyclophylla* dwarf-shrublands colonize low-lying areas where the soils are siltier. The higher, sandy areas on the periphery of these plains are occupied by *Hammada salicornica* dwarf-shrublands.
Further west, on the sands of Sahl Rakba (= Rakba plain), the *Salsola cyclophylla* community covers large areas. It thus makes up a large portion of the dwarf-shrub vegetation, which is physiognomically dominant especially in the peripheral parts of this plain.
In the study area dwarf-shrublands characterized by *Indigofera spinosa* are limited to Sahl Rakba, the eastern edge of which is also their eastern limit of distribution. Growing on a mostly sandy substrate, this unit occurs extensively as well as in islandlike stands. Mixed stands of *Indigofera spinosa* and *Panicum turgidum* or *Salsola cyclophylla* also occur, whereas according to our observations *Panicum turgidum* and *Salsola cyclophylla* exclude one another for the most part.
It still is not known what climatic factor is responsible for the restriction of *Indigofera spinosa* to the western part of the study area.
In Sahl Ḥadaba there is a dwarf-shrub community characterized by *Traganum nudatum* which occurs in small patches on islands of sand over very compact and very saline soils (Sample 26, Table 2).

Table 2 Dwarf-scrub and grasslands of the Arabian Shield (western and central Nağd). Extract of table 5 in BAIERLE/EL-SHEIKH/FREY (1985).

Serial number :	1	2	3	4	5	6	7	8	9	10	11	12	13	14	15	16	17	18	19	20	21	22	23	24	25	26
Number of species:	15	14	17	10	21	15	16	7	4	2	1	2	4	3	2	2	7	3	5	4	3	2	4	1	2	2
Character species:																										
Hammada salicornica	1	2a	1	1	2a	1	2a	2a	2a	1	1	1	+										(+)			
Salsola cyclophylla								r				+	1	1	1	2a	1	1	r	+	+					(+)
Indigofera spinosa																r	1	2m	2m	r						
Panicum turgidum									+	r			r	r	(+)				1	1	2a	1	2a	1	+	1
Lasiurus scindicus								+															+		2a	
Traganum nudatum																										2a

Geographic location of samples:

Serial numbers	Location
1–26	Arabian shield (western and central Nağd)
1–9	East of Ğabal Ẓalim
10–26	West of Ğabal Ẓalim
10–15	East of Rakba-plain
16–21	Rakba-plain
22–26	East of Rakba-plain (between Ğabal Ẓalim and al Muwaih)

Edaphic characteristics:

Serial numbers	Characteristics
1–9	Sand and fine gravel (mainly granit sands) moderately saline (intensified Na- and K-contents)
10–11	Gyps., high Na- and Cl-values
12–26	Mainly sands
20–26	Relatively thin sand layer on gravel
12–24	Not saline to moderately saline
25–26	High Na- and Cl-val.

Locality of samples (to table 2)
Locality, sea level

Sample			
1	(63):	11 km WSW ad-Dawādimī;	1050 m above sea-level
2	(64):	22 km WSW ad-Dawādimī;	980 m above sea-level
3	(59):	23 km NE ad-Dawādimī;	900 m above sea-level
4	(60):	23 km NE ad-Dawādimī;	900 m above sea-level
5	(62):	17 km NNW ad-Dawādimī;	950 m above sea-level
6	(78):	22 km SSW 'Afīf;	1020 m above sea-level
7	(79):	22 km SSW 'Afīf;	1020 m above sea-level
8	(83):	79 km SW 'Afīf;	1000 m above sea-level
9	(84):	79 km SW 'Afīf;	1000 m above sea-level
10	(95):	18 km WSW al-Muwaih;	950 m above sea-level
11	(96):	18 km WSW al-Muwaih;	950 m above sea-level
12	(94):	14 km WSW al-Muwaih;	950 m above sea-level
13	(93):	14 km WSW al-Muwaih;	950 m above sea-level
14	(92):	5 km WSW al-Muwaih;	950 m above sea-level
15	(91):	5 km WSW al-Muwaih;	950 m above sea-level
16	(108):	36 km NE aṭ-Ṭā'if;	1320 m above sea-level
17	(107):	70 km SW Riḍwān;	1260 m above sea-level
18	(106):	60 km SW Riḍwān;	1190 m above sea-level
19	(104):	32 km SW Riḍwān;	1060 m above sea-level
20	(100):	25 km NE Riḍwān;	960 m above sea-level
21	(101):	25 km NE Riḍwān;	960 m above sea-level
22	(90):	42 km SW Ẓalim;	
23	(85):	17 km NE Ẓalim;	1000 m above sea-level
24	(86):	17 km NE Ẓalim;	1000 m above sea-level
25	(89):	29 km SW Ẓalim;	990 m above sea-level
26	(88):	12 km SW Ẓalim;	980 m above sea-level

All samples recorded except 10/11 (tertiary gypsum strata) and 20/21 (tertiary basalts) are situated above different precambrian formations of the crystalline above loose sediments (sands, gravels) of various thickness. The large-scale loose sediment areas (for instance Rakba-plain) are entered in the geological map as Qu-formation.

Extensive grasslands are found only in the vicinity of Ẓalim, where stands of *Panicum turgidum*, usually severely overgrazed, colonize shallow to moderately deep layers of sand and fine gravel at the bottoms of slopes and near Ǧabal Ẓalim (Samples 23 and 24, Table 2, Photograph 3). Small areas of *Panicum turgidum* and *Lasiurus scindicus* grassland also occur west of Ẓalim in slightly to moderately saline sand islands and depressions in the pebble and gravel plains of Sahl Ḥadaba. Other stands are found on the eastern margin of Sahl Rakba.

At present it cannot be said whether the conspicuous paucity of species that was observed in the dwarf-shrub communities and grasslands west of Ẓalim was a result of precipitation conditions during the year of the investigation or whether other factors are responsible.

4.1.3 Wadi communities

The small wadis, which are generally free of trees and shrubs, also have predominantly dwarf-shrub vegetation. The water supply in these areas is relatively good on account of drainage from surrounding regs or from major catchment areas. It must therefore be concluded that trees and shrubs were once present in some of the larger wadis but were removed by man. The physiognomic dominance of dwarf-shrubs is thus partially a secondary phenomenon.

In wadis of the cuesta region and the eastern Arabian Shield three plant communities could be distinguished. The primary factor determining their distribution is water supply, and closely related to this are the depth of the soil and its physical and chemical characteristics. The small, rocky channels containing little fine earth that drain the regs characteristically have a high coverage of *Cymbopogon commutatus*, which belongs to the Poaceae (Photograph 4). In wadis and channels where there is a thicker but often discontinuous cover of sandy and sandy-silty fine material, the characteristic species is *Rhanterium epapposum*. Dwarf-shrublands dominated by *Zygophyllum coccineum* are found in the cuesta region on hardened clayey and lime-rich wadi soils that form a closed cover of fine material.

These three characteristic species are frequently accompanied by *Anvillea garcinii*, *Gymnocarpos decandrum*, *Zilla spinosa* and *Fagonia bruguieri*. In the channels and channel segments rich in *Cymbopogon commutatus* these latter species predominate, and they also have the highest coverage. They all are character species of the Anabasetea articulatae Zoh. Their occurrence also clearly distinguishes the *Rhanterium epapposum* wadi community from the extensive *Rhanterium epapposum* dwarf-shrublands, which occur in exclusively sandy habitats. Other companions are *Helianthemum lippii*, *Salvia aegyptiaca* and *Farsetia aegyptiaca* (wadis poor in fine soil), as well as *Astragalus spinosus*, *Heliotropium ramosissimum* and *Lasiurus scindicus* (unbroken coverings of fine material).

In the wadis of the cuesta region and the eastern Arabian Shield there are other communities dominated by dwarf-shrubs or grasses that were not recorded in the samples. These are the *Hammada salicornica* dwarf-shrublands (occurring on thick, moderately saline sand covers), *Panicum turgidum* grasslands, *Pituranthos triradiatus* and *Pulicaria crispa* dwarf-shrublands, and a substitute community characterized by *Rhazya stricta*. *Rhazya stricta* tends to be dominant especially in vegetation units which have been severely disturbed. In the vicinity of settlements (e.g. Šaqrā' and Marāh) it can replace the original vegetation over wide areas. It is not yet known to what extent the spread of the *Zygophyllum coccineum* community is likewise promoted by grazing pressure in the wadis (according to Kassas [1953] *Zygophyllum coccineum* is not eaten by grazing animals).

No surveys were conducted in the small wadis of the central and western Arabian Shield. The vegetation of the treeless and shrubless wadis there probably differs from that described above much in the way that the undergrowth of the scrub and of the pseudo-savannas is different in the eastern and the western part of the transect (cf. Section 4.2.2).

4.2 Woodlands and scrub

In central Saudi Arabia vegetation units whose physiognomy is characterized by arboreal species cover large areas only on the al-Ḥiǧāz slopes; small stands are found in wadis throughout the study area. The following division of the woodland units into two groups reflects floristic differences and contrasting habitat requirements.

Scrub communities are also described here in view of the alterations in vegetation structure which one can observe to be taking place, especially in the wadis, as a result of human interference.

No attention is given here to the extremely open stands of *Acacia tortilis* which are occasionally found east of al-Ḥiǧāz outside wadis and alluvial plains.

4.2.1 Woodlands on the slopes of the southern al-Ḥiǧāz

In most of the study area the characteristic vegetation units are dwarf-shrublands and grasslands, with intervening expanses of desert; extensive units whose physiognomy is determined by arboreal species are confined to the western part of the study area. Here *Extremely xeromorphic woodlands* containing various *Acacia* species cover large areas of the rocky al-Ḥiǧāz slopes.

The composition of the extremely patchy tree layer (Photograph 5) is determined mainly by the altitude of the location, the distance from the watershed, the effects of exposure and the nature of the subsoil.

North of aṭ-Ṭā'if, in the uppermost zone of the *Thorn-woodlands* between 1350 and 1500 m, *Acacia asak* and *A. etbaica* are dominant. The *Thorn-woodlands* belong to the class Acacietea sudano-arabica Zoh. At lower and dryer sites these species are replaced by *Acacia tortilis* and *Acacia hamulosa. Lycium shawii* is often found in the shrub layer.

The undergrowth, dominated mainly by dwarf-shrubs, also has many gaps and seldom covers more than 10% of the stony substrate, which is poor in fine soil. Frequent species (constancy $\geq$ 50%) are *Justicia odora, Hibiscus micranthus, Indigofera spinosa, Fagonia indica, Aerva persica, A. lanata, Kickxia spartioides, Commicarpus helenae, C. sinuatus, Salsola cyclophylla, Tetrapogon villosus, Solanum albicaule, Blepharis ciliaris, Otostegia fruticosa* and *Aristida adscensionis.*

Given comparable degrees of coverage the *Thorn-woodlands* on the slopes play different parts in the landscape depending on the subsoil. On mountain tops, which are built up of huge granite blocks, the trees and shrubs are often concealed among the rock masses, while on slopes, which consist of slaty rock, loose "orchardlike" stands develop.

As one goes further east from the watershed, the *Thorn-woodlands* of al-Ḥiǧāz diminish. In a lava field near 'Ušaira (60 km NNE of aṭ-Ṭā'if), for example, the only species still found of those mentioned above were *Hibiscus micranthus*, *Indigofera spinosa*, *Salsola cyclophylla*, *Solanum albicaule* and *Blepharis ciliaris*. Although the altitude here is 1400 m, *Acacia asak* and *A. etbaica* are absent. In the dark basalt hammada at this location there are only a few scattered individuals of *Acacia hamulosa* and *A. tortilis*.

4.2.2 Pseudo-savannas and scrub in wadis and other alluvial stands (Table 3)

East of al-Ḥiǧāz, woodlands and scrub are limited to wadis and other habitats where groundwater is near the surface. After rains large quantities of water often flow through the wadis, moistening the alluvial soils to a fairly good depth. The moisture stored in the sediments, along with the groundwater, makes it possible for bushes and trees to grow in a narrow zone in the central parts of wadis.

Here again the tree and shrub layers of the vegetation units are characterized mainly by *Acacia* species. *Tamarix* species were not observed at the locations studied, and *Ziziphus spina-christi* was found only in one wadi near 'Ušaira (cf. GIACOMINI et al. 1979a, b and VESEY-FITZGERALD 1957b).

The wadi communities, most of which belong to the class Acacietea tortilis sub-sudanica ZOH. 1973, have *Acacia raddiana* and *A. tortilis* as their most important species. *Acacia tortilis* apparently requires less moisture than *A. raddiana*, since it also occurs in the smaller and drier wadis of the cuesta region (Samples 2 and 3, Table 3) and at locations in Sahl Rakba that receive very little run-in water (Samples 20–22, Table 3). *Acacia gerrardii* is of less importance. However, this species is typical of the pseudo-savannas in the wadis of the al-Ḥiǧāz region. The dominant shrubs are *Lycium shawii* and *Acacia ehrenbergiana*. In addition, though, shrublike individuals of *Acacia tortilis*, *A. raddiana* and *A. gerrardii* make up a considerable part of the wadi scrub.

It can be assumed that an original natural distribution of *Acacia* species, determined by factors such as climatic differences and variations in water supply (cf. HALEVY/ORSHAN 1972), has been altered by anthropozoogenic influences to the point where it can no longer be discerned. This hypothesis is supported by the fact that even in the large wadis there are almost no acacias with trunks in the densely settled cuesta region. For this reason the distribution of tree and shrub species will not be further discussed here.

The scrub and woodland units of Sahl Rakba, which are probably mainly dependent on groundwater, show great floristic similarity to the wadi communities of the same region and are therefore listed together with these in Table 3.

Dwarf-shrub and herb layer

Whereas in the tree and shrub layers at least four of the important species occur throughout the study area, namely *Acacia tortilis*, *A. gerrardii*, *A. ehrenbergiana* and *Lycium shawii*, in the undergrowth there is a clear geographical division of the vegetation units into a western and an eastern complex of communities.

In the wadi scrub and pseudo-savannas of the western region (western and central Arabian Shield between aṭ-Ṭā'if and Ẓalim, including the units of Sahl Rakba, which are probably dependent on groundwater) the undergrowth is extremely poor in species, consisting almost exclusively of *Panicum turgidum*, *Salsola cyclophylla* and *Indigofera spinosa*. In contrast, the dwarf-scrub and herb layer of the eastern Arabian Shield and the Arabian cuesta region has a much greater number of species and is more highly differentiated.

As was the case in section 4.1.3 (wadi communities rich in dwarf-shrubs), the properties of the substrate have an important influence on the composition of this layer. Tussock grasses such as *Hyparrhenia hirta*, *Chrysopogon plumulosus* and *Cymbopogon commutatus* characterize the smallest runnels, which have little fine soil (Samples 1–3, Table 3). *Rhanterium epapposum* is typical of most of the wadi soils in the intermediate ecological range (Samples 4 and 5 and 8 to 14, Table 3), while *Hammada salicornica* is restricted to moderately saline sandy soils (Sample 17, Table 3).

Like the shrub and tree layers, the undergrowth of the wadi scrub and pseudo-savannas shows the effects of anthropozoogenic influences on the vegetation, especially heavy grazing. This is even more evident than in the wadi dwarf-scrub communities. One sign of these influences is the high species magnitude attained by the poisonous *Rhazya stricta* (Apocynaceae) and the thorny *Zilla spinosa* (Brassicaceae) in some wadis (Samples 6 and 7; Samples 7, 9, 10 and 14; Table 3). Another sign is the fact that two species, *Astragalus spinosus* and *Zilla spinosa*, occur with a maximum degree of constancy in all of the units. On account of their thorny habit they are seldom or never eaten by animals (cf. also THALEN 1979 and MCGINNIES et al. 1968).

In Table 3 the samples in which the vegetation is characterized by trees are indicated by an asterisk (*).

4.3 Vegetation on sand dunes (Nafūd as-Sirr, Nafūd Qunaifiḏa)

The vegetation on sand dunes varies considerably depending on their degree of consolidation, topography and moisture conditions. There are dunes with permanent vegetation, with no vegetation, and with a cover of annuals after winter rains.

On dunes with perennial vegetation the shrub *Calligonum comosum*, growing to a height of about 1 m, is dominant. The vegetation cover is irregular and the individual plants are often widely separated (Photograph 7). Other species of the *Calligonum comosum* community which are characteristic, although not so physiognomically conspicuous, are *Cyperus conglomeratus*, *Moltkiopsis ciliata*, *Danthonia forsskalii*, *Dipcadi erythraeum*, *Panicum turgidum*, *Stipagrostis plumosa*, *Polycarpaea repens* and *Rhanterium epapposum*. While the last four species named are also common in other vegetation units or are character species of other communities, *Cyperus conglomeratus*, *Moltkiopsis ciliata*, *Danthonia forsskalii* and *Dipcadi erythraeum* clearly have their focus of distribution in these dune areas. Like *Calligonum comosum* they are character species of the class Retametea retami ZOH. *Artemisia monosperma*, which was described by VESEY-FITZGERALD (1957 b) and GIACOMINI et al. (1979 a) as a characteristic component of the dune vegetation in parts of the ad-Dahnā', the Nafūd and the ar-Rub' al-Ḫālī (often co-dominant with *Calligonum comosum*), was not observed in the area studied.

The occurrence of annuals is related to precipitation and to the topography of dunes. On the crests and many of the slopes the vegetation consists almost exclusively of perennial species. If there was sufficient rainfall annuals are found predominantly at the bases of dunes and in hollows or other depressions. Here the water content of the upper soil layers, which is essential for annuals, is much greater than in parts of dunes that are higher and are exposed to the wind. This is because of surface run-in water, emerging seepage water and slower evaporation by wind.

Frequent species in these annual meadows are *Eremobium lineare*, *Astragalus schimperi*, *Astragalus gyzensis*, *Plantago cylindrica* and *Neurada procumbens*. These species are not limited to the sand dune region. They also occur as companions in many of the vegetation units of the study area when conditions in the upper soil layers are favourable, i. e. when there is a thin, loose, moist layer of sand.

The coverage in such annual "meadows" can reach relatively high levels. In one hollow, for example, it amounted to 35%. The landscape, normally yellow, thus becomes green for a brief period.

Water conditions in the dunes are generally favourable for plants such as *Calligonum comosum*, whose long root systems can reach down to the water stored in deeper soil layers: Penetration of rainwater into the sand is good and the moisture becomes 'sealed in' when the surface layers dry out (VESEY-FITZGERALD 1957b). Even so, in many places *Calligonum comosum* and other accompanying perennial species are almost entirely absent. The cause of this is probably to be found in anthropozoogenic influences such as grazing and cutting of shrubs for fuel, as well as a lack of consolidation in these parts of the dunes.

4.4 Deserts, vegetation on rock desert plains

Large parts of central Saudi Arabia are taken up by mostly sterile sand, pebble, gravel, rock and salt deserts. Many of these desert areas have annual "meadows" only after sufficient spring rains.

In parts of the regs in the cuesta region there are small numbers of individuals of the species *Gymnocarpos decandrum*, *Hammada salicornica*, *Halothamnus bottae*, *Rhanterium epapposum*, *Anvillea garcinii*, *Farsetia aegyptiaca*, *Fagonia bruguieri* and *Stipagrostis plumosa*. These are elements of the neighbouring dwarf-scrub communities. They colonize the regs when there is some kind of sand cover – even if it has many gaps – on the boulder pavement. The only species of the above mentioned which is also typical of bare rock surfaces is *Fagonia bruguieri*.

In contrast, other parts of the regs are characterized by vegetation with a very typical species composition. One finds a physiognomically inconspicuous, sparse vegetation consisting of small, 5- to 10-cm-high tussocks of various *Oropetium* species (*O. africanum*, *O. capense* and *O. capense* var. *arabicum*; cf. SCHOLZ/KÖNIG 1984), of *Anastatica hierochuntica* or of *Blepharis ciliaris*. It remains to be determined which factors are responsible in each case for the occurrence or absence of these species. The geology of the subsoil is apparently not of particular importance in this regard. The above-named species occur both in the regs of the cuesta region and in various plains of the crystalline zone.

In true rock deserts the perennial vegetation that determines the physiognomy is limited to drainage channels and wadis, which are often only a few metres wide.

5. Summary

Based on the evaluation of 115 vegetation samples a description is given of the vegetation units occurring in the vicinity of the transect aṭ-Ṭā'if–ar-Riyāḍ (central Saudi Arabia):

- dwarf-scrub and related communities, including dwarf-scrub vegetation of the wadis (these are the prevailing formations in the study region with respect to area covered),
- *Panicum turgidum* grasslands,
- *Thorn-woodlands* of the class Acacietea sudano-arabica Zoh. on the eastern al-Ḥiǧāz slopes,
- pseudo-savannas and shrub formations of the wadis (Acacietea tortilis sub-sudanica Zoh.), which are dependent on groundwater and run-in water,
- the vegetation of the dune areas in the cuesta region,
- the characteristic sparse vegetation of the regs.

The distribution patterns of the communities and groups of communities are analysed. For the dwarf-scrub, the pseudo-savannas, the wadi scrub and other formations, geographically clearly defined units are identified.
The vegetation conditions described are documented by means of photographs and three survey tables.

Zusammenfassung

Aufbauend auf der Auswertung von 115 Vegetationsaufnahmen werden die im Bereich des Transektes aṭ-Ṭā'if–ar-Riyāḍ (Mittleres Saudi Arabien) vorkommenden Vegetationseinheiten beschrieben:

- die flächenmäßig im Untersuchungsgebiet vorherrschenden Zwergstrauchreichen Formationen einschließlich der Zwergstrauchreichen Vegetation der Wadis,
- *Panicum turgidum*-Grasfluren,
- *Dorn-Offenwälder* der Klasse Acacietea sudano-arabica Zoh. der al-Ḥiǧāz-Ostabhänge,
- Pseudo-Savannen und Gebüschformationen der Wadis (Acacietea tortilis sub-sudanica Zoh.), die auf Grund- und Zuschußwasser angewiesen sind,
- die Vegetation der Sanddünengebiete im Schichtstufenland,
- der charakteristische schüttere Bewuchs der Reg-Flächen.

Die Verbreitungsmuster der vorkommenden Gesellschaften bzw. Gesellschaftsgruppen werden analysiert. Hierbei ergeben sich u.a. bei den Zwerggesträuchen, den Pseudo-Savannen und den Gebüschen der Wadis geographisch klar definierbare Einheiten.
Die dargestellten Vegetationsverhältnisse werden durch drei Übersichtstabellen und Photos dokumentiert.

6. Literature

Al-Sayari, S. S. / Zötl, J. G. (eds.), Quarternary period in Saudi Arabia. 1. Sedimentological, hydrogeological, geomorphological and climatological investigations in Central and Eastern Arabia. 1978, Wien.

Baierle, H. U. / El-Sheikh, A. M. / Frey, W., Vegetation und Flora im mittleren Saudi-Arabien (aṭ-Ṭā'if–ar-Riyāḍ). Beih. Tübinger Atlas Vorderer Orient, Reihe A, Naturwissenschaften, *22*, 1985, Wiesbaden.

Baierle, H. U. / El-Sheikh, A. M. / Frey, W. / König, P. / Kürschner, H., Mittleres Saudi-Arabien. Vegetation. Karte A VI 9 des Tübinger Atlas des Vorderen Orients, 1985, Wiesbaden.

Barth, H. K., Landesnatur. In: Blume, H. (Hrsg.), Saudi-Arabien – Natur, Geschichte, Mensch und Wirtschaft, pp. 23–88, 1976, Tübingen, Basel.

Batanouny, K. H., Vegetation along Jeddah–Mecca road: Pattern and process as affected by human impact. Journ. Arid Environm. *2*, 1979, 21–30.

Batanouny, K. H. / Baeshin, N. A., Plant communities along the Medina–Bahr road across the Hejaz mountains, Saudi Arabia. Vegetatio *53*, 1983, 33–43.

Brown, G. F., Geomorphology of Western and Central Saudi Arabia. 21st Int. Geol. Congr. Copenhagen, Rep. *21*, 1960, 150–159.

De Marco, G. / Dinelli, A., First contribution to the floristic knowledge of Saudi Arabia. Ann. Bot. *33*, 1974, 209–236.

El-Wailly, A. J., Seasonal variation in desert flora at Riyadh, Saudi Arabia. Bull. Iraq Nat. Hist. Mus. *5* (2), 1972, 10–16.

Frey, W. / Probst, W., Gliederung der Vegetation und ihre Darstellung im Tübinger Atlas des Vorderen Orients (TAVO) und in den Beiheften zum Atlas. Beihefte Tübinger Atlas Vorderer Orient, Reihe A (Naturwissenschaften), *1*, 1977, Wiesbaden.

Geological Maps of the Kingdom of Saudi Arabia, 1 : 500 000:
I – 206 A (Wadi Ar Rimah Quadrangle)
GM – 207 A (Northern Tuwayq Quadrangle)
I – 210 A (Southern Hijaz Quadrangle)
GM – 211 A (Southern Najd Quadrangle)
U.S. Geological Survey, Washington D.C. (1958 ff.)

Giacomini, V. / Longhitano, N. / Corti, L., Cartography of the vegetation of the Eastern (Hasa) Province of Saudi Arabia. Publ. Ist. Bot. Univ. Catania 1979, *1*, 1979 a, 1–56.

Giacomini, V. / Longhitano, N. / Corti, L., The vegetation and grazing resources of Wadi Al Miyah area, El Hasa Province of Saudi Arabia. Publ. Ist. Bot. Univ. Catania 1979, *1*, 1979 b, 1–50.

Halevy, G. / Orshan, G., Ecological studies on *Acacia* species in the Negev and Sinai. I. Distribution of *Acacia raddiana*, *A. tortilis* and *A. gerrardii* ssp. *negevensis* as related to environmental factors. Israel Journ. Bot. *21*, 1972, 197–208.

Kassas, M., Habitat and plant communities in the Egyptian desert. II. The features of a desert community. Journ. Ecol. *41*, 1953, 248–256.

Mandaville, J. P., Notes on the vegetation of Wadi As-Sahba, Eastern Arabia. Journ. Bombay Nat. Hist. Soc. *62*, 1965, 330–332.

McGinnies, W. G. / Goldman, B. J. / Paylore, P. (eds.), Deserts of the World 1968, Tucson.

Meteorologische Daten (Monthly Data, Hydrological Publications) aus dem Beobachtungszeitraum 1967–1976 (unvollständig). Hrsg.: Kingdom of Saudi Arabia, Ministry of Agriculture and Water, Dept. of Water Resources Development, Hydrology Division, Riyadh.

Migahid, A. M., Flora of Saudi Arabia. 1978, 2nd ed., 2 vols., Riyadh.

Migahid, A. M. / El-Sheikh, A. M., Types of desert habitat and their vegetation in Central and Eastern Arabia. Proc. Saudi Biol. Soc. *1*, 1977, 5–13.

Powers, R. W. / Ramirez, L. F. / Redmond, C. D. / Elberg, E. L. Jr., Geology of the Arabian Peninsula – Sedimentary geology of Saudi-Arabia. U.S. Geological Survey Professional Paper No. 560 D, 1966, Washington.

Scholz, H. / König, P., Die Gattung *Oropetium* (Gramineae) in Arabien. Willdenowia *14*, 1984, 159–163.

Thalen, C. D. P., Ecology and utilization of desert shrub rangelands in Iraq, 1979, The Hague.

Topographische Karten: Arabian Peninsula: 1:500000 (Series K462, Edition 1-AMS), sheets NF 37-E (Southern Hijaz), NF 38-W (Southern Najd) and NG 38-W (Wadi Ar Rimah). Hrsg.: Army Map Service (AM), Corps of Engineers, U.S. Army, Washington D.C. (1963).

Vesey-Fitzgerald, D. F., The vegetation of the Red Sea coast north of Jeddah, Saudi Arabia. Journ. Ecol. *45*, 1957 a, 547–562.

Vesey-Fitzgerald, D. F., The vegetation of Central and Eastern Arabia. Journ. Ecol. *45*, 1957 b, 779–798.

Zohary, M., Geobotanical foundations of the Middle East. 1973, 2 vols., Stuttgart.

Zohary, M., Vegetation of Israel and adjacent areas. Beih. Tübinger Atlas Vorderer Orient, Reihe A (Naturwissenschaften), *7*, 1982, Wiesbaden.

Dipl.-Biol. Heinz Ullrich Baierle, Prof. Dr. Wolfgang Frey, Institut für Systematische Botanik und Pflanzengeographie, Freie Universität Berlin, Altensteinstr. 6, D-1000 Berlin 33.

Photo 1 *Rhanterium epapposum*-dwarf-shrubland on closed sand cover 20 km ESE Ḫuff.

Photo 2 *Hammada salicornica* dwarf-shrubland 10 km west of ad-Dawādimī on granite sand.

Photo 3 *Panicum turgidum*-stands near the Ǧabal Ẓalim east of Ẓalim.

Photo 4 *Cymbopogon commutatus* rich association in a small wadi on reg plains of the Dhrumah formation 20 km northwest al-Barra; in the foreground *Rhanterium epapposum*.

Photo 5 Very open xeromorphic *Acacia*-woodland on the al-Ḥiǧāz slopes 30 km NNE of aṭ-Ṭā'if.

Photo 6 Pseudo-savanna near Bi'r Siǧā, extremely devastated by overgrazing.

Photo 7 *Calligonum comosum*-association on stabilized sand dunes of the Nafūd as-Sirr.

Photo 8 Rock desert 20 km northwest al-Barra (Dhrumah-formation).

Peter König

Zonation of vegetation in the mountainous region of south-western Saudi Arabia (ʿAsīr, Tihāma)*

Contents

1. Introduction

Up to the end of the 1970s our knowledge of vegetation conditions in south-western Saudi Arabia was based largely on the studies of Vesey-Fitzgerald (1955, 1957 a, b) and Popov/Zeller (1963). These authors carried out investigations in zoology as well as botany. Wissmann (1972) concerned himself with the '*Juniperus*-Gebirgswälder' of Arabia.
In recent years studies have been published on subregions. For example, Batanouny (1979), Batanouny/Baeshin (1983) and Baierle et al. (1985) have dealt with the northern part of the study area. The southern part has been reported on by Abulfatih (1979, 1981 a) and Brooks/Mandil (1983).

---

* Part of a dissertation (König, in prep.).

Clues about the distribution of a number of species and thus about their zonation patterns can be obtained from accounts of journeys such as that by THESIGER (1947) or from studies with a floristic orientation (SCHWARTZ 1939, KHATTAB/EL-HADIDI 1971, MANDAVILLE 1973, BATANOUNY/BAESHIN 1978, ABULFATIH 1981b, 1984).

In this study profiles are used to present some recent ideas about the zonation and structure of vegetation in the mountainous and hilly regions of ʿAsīr and Tihāma in south-western Saudi Arabia.

The exploring expeditions done in 1981 and 1982 were conducted with support from the Sonderforschungsbereich 19 'Tübinger Atlas des Vorderen Orients' / Deutsche Forschungsgemeinschaft and the Saudi Biological Society / Saudi Arabian National Center for Science and Technology, for which I would like to express my particular thanks.

I am indebted to H. U. Baierle, W. Frey and H. Kürschner (Berlin) as well as to S. A. Chaudhary, A. M. El-Sheikh, H. M. Hassan, A. Mahmoud and M. Odat (ar-Riyāḍ) for their help in the field work.

I am grateful to A. J. C. Grierson, I. C. Hedge, R. A. King, J. Lamond and A. G. Miller (Edinburgh), to H. Scholz (Berlin) and to A. Hunde (Uppsala) for their help in determination and revision of critical taxa.

The collected plant specimens are in the 'Herbarium Vorderer Orient' of the Institut für Systematische Botanik und Pflanzengeographie, Freie Universität Berlin (BSB). Duplicates – where existent – are in the herbaria of the Botany Department, Faculty of Science, King Saud University, ar-Riyāḍ and the Royal Botanic Garden, Edinburgh (E).

2. Geography and geology

Together with the mountain regions of the Yemen Arab Republic and the People's Democratic Republic of Yemen, the mountain regions of south-western Saudi Arabia (ʿAsīr, hilly Tihāma) (Fig. 1) make up what is known as 'Arabia felix'. The study area extends from Ǧidda and aṭ-Ṭā'if in the north (22° N, 39–41° E) to the border with the Yemen Arab Republic in the south (Ǧīzān, Abhā and Naǧrān) (17° N, 42–44° E).

The region is characterized by a high relief intensity: 0 m on the Red Sea to more than 3000 m on Ǧabal Saudā', which is north-west of Abhā and is known for its national park (Fig. 2a).

The Tihāma and ʿAsīr are distinct natural regions. The Tihāma extends from the Red Sea coast to the escarpment of the border range and can be divided into the Tihāma coastal plain and the hilly Tihāma (Fig. 2b). Whereas the former is characterized by broad plains with a sandy-gravelly subsoil and by elevations up to 100(–300) m, the hilly Tihāma comprises

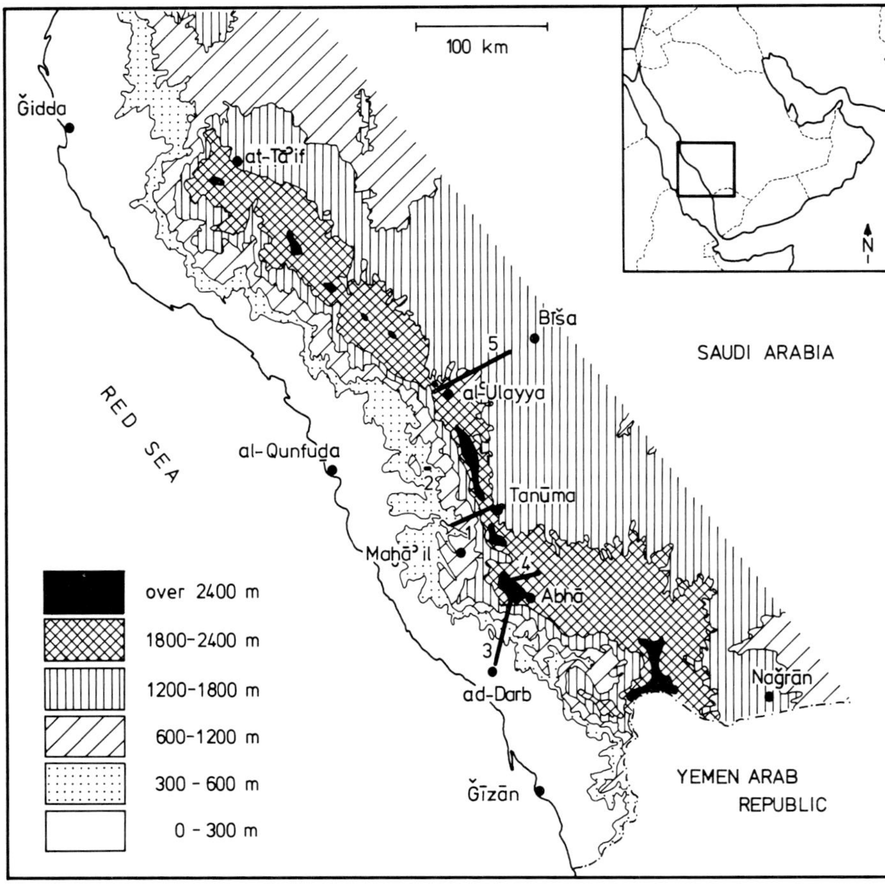

Fig. 1 Geographical location and orography of the study area, and location of vegetation profiles in SW Saudi Arabia.

the hill zone and the border range up to an altitude of about 800 m along with isolated mountains in front of the escarpment.

The highlands, including the escarpment of the range bordering the Red Sea and the eastern slopes, are known as ʿAsīr and occasionally as al-Ḥiğāz. In the present study ʿAsīr will be understood as the region between aṭ-Ṭā'if in the north and the border with the Yemen Arab Republic in the south. North of aṭ-Ṭā'if the mountains of al-Ḥiğāz begin.

Geologically and structurally the south-western part of Arabia belongs to the Arabian Shield, which separated in the late Tertiary from the African continent when the Red Sea and Gulf of Aden rifts opened and joined the East African Rift Valley system (triple junction). The Precambrian basement, a truncated upland, is made up mostly of granites, gneisses and highly metamorphosed, crystalline slates. The uplifting of the rift margins

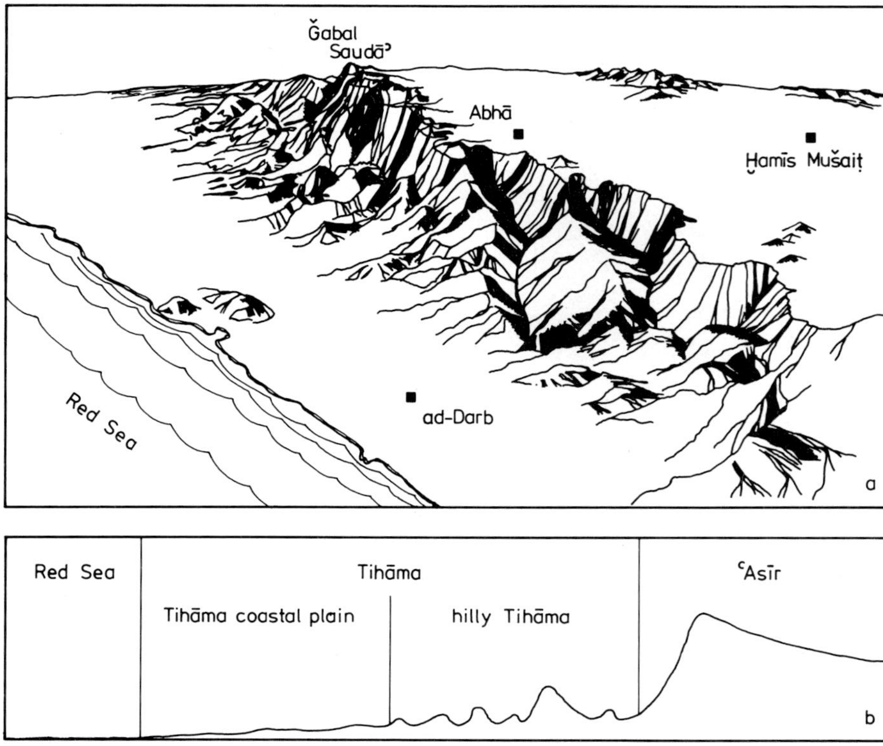

Fig. 2 Structural shape in the southern part of the study area near Abhā (illustrated according to a diagram of the "ʿAsīr National Park" administration) (a), and schematic diagram of the geographical units (b).

that took place with the spreading of the rift produced the pronounced escarpment along the Red Sea coast.

In addition, as a result of volcanic activity in the faulting which accompanied rift spreading, basaltic and rhyolithic magmas welled up and formed extensive volcanic rock deserts (harrah, harrat), overlying the basement in some places.

In the Tihāma coastal plain and along the eastern mountain slopes near aṭ-Ṭāʾif and Naǧrān, Quaternary gravels and sands predominate. There are also dunes, the most extensive formations of which are in the ar-Rubʿ al-Ḫālī desert area.

Additional information and literature references can be found in BARTH (1976), CHAPMAN (1978) and the GEOLOGICAL MAP OF THE KINGDOM OF SAUDI ARABIA 1 : 500 000 (1958–79), from which the information in this section has been obtained.

3. Climate

The large-scale distribution and zonation of vegetation units is determined mainly by climatic conditions.

Of decisive importance is the seasonal succession of precipitation (winter and summer rains). There exist two wind systems. On the one hand there are the winter cyclones of the west-wind zone (northern winds), which are responsible for the winter and spring rains in the northern parts of the country and which shift to the south in the winter and affect the southern regions. On the other hand there are monsoon winds from the south-west that bring additional summer precipitation to the south-western regions. The northern limit of the area thus influenced is situated roughly at the latitude of Makka (Mecca) (cf. Fig. 3) (Rathjens et al. 1956, Barth 1976).

According to Rathjens et al. (1956) the northern winds have a vertical depth of about 3000 m and adhere to the surface relief. In contrast, the monsoon winds exert an influence only up to a maximum of 2000 m. They are therefore obstructed and deflected by the south-western border range, and orographic precipitation occurs (fog zone, condensation zone).

According to Flohn (1965) these effects are enhanced by the diurnal wind system, which develops between the sea and the mountains on each side of the Red Sea (indigenous Red Sea winds) and creates a condensation zone at 1000–2500 m. Here there is often fog and high humidity.

These zones of high humidity are characterized by 'cloud forests' (here mainly classified as sclerophyllous scrub and *Juniperus* woodland), which have been described by Troll (1935) and Kassas (1956) for the Red Sea Hills of the Sudan, by Troll/Schottenloher (1939) for Eritrea and by Rathjens et al. (1956) for ʿAsīr and Yemen.

A general survey of climatic conditions is given by the map of climatic diagrams (Fig. 3), which is based on data from meteorological stations.[1] The diagrams were prepared according to the method of Walter (1973). The transect through the stations al-Mud̲ailif (Al Mudhaylif), Miḫwā (Mikhwa), Balǧurašī (Baljarshi) and Bīša (Bishah) clearly shows how climatic conditions are dependent on relief and altitude. The sharp drop in precipitation values east of the escarpment is illustrated by the stations Saudā' (Sawdah), Abhā (Abha) and Ḫamīs Mušaiṭ (Khamis Mushayt), which lie close to one another.

The northern boundary of the summer-rain zone is located approximately at the latitude of Makka. This is indicated by small peaks in August in the climatic diagram of as-Sail al-Kabīr (As Sayl al Kabir) and al-Ḥawīya (Al Hawiyah), which lie in the transition zone between ʿAsīr and al-Ḥiǧāz. As

---

[1] Hydrological Publications, Ministry of Agriculture and Water, Department of Water Resources Development, Hydrology Division, ar-Riyāḍ.

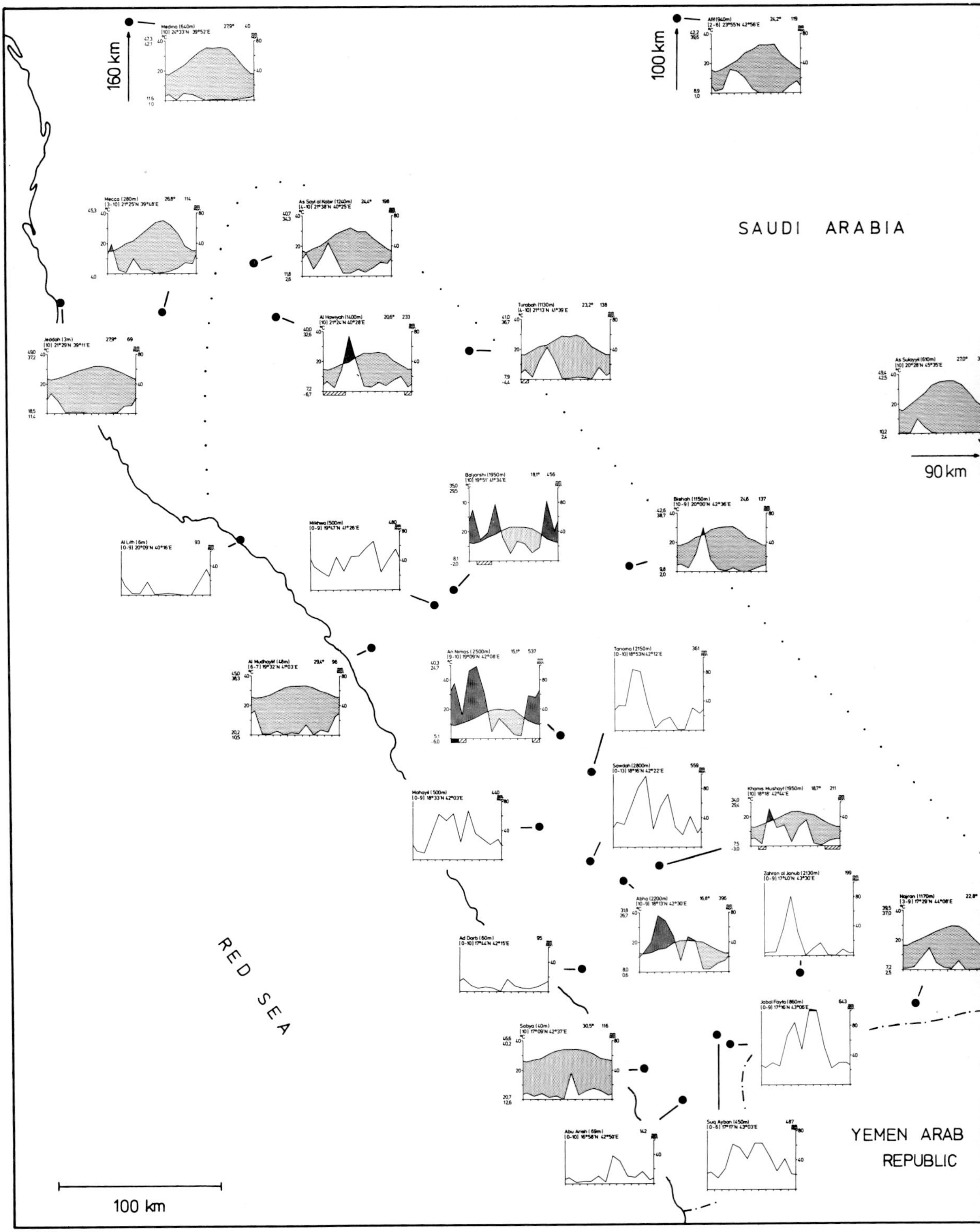

Fig. 3 Map of climatic diagrams of SW Saudi Arabia; dotted line indicates the northern boundary of summer rains. al-Madīna, ʿAfīf and as-Sulayyil are not to scale in the distance to the other stations.

can be seen from the diagram of al-Madīna (Medina), located more than 300 km north of Makka, there is no summer precipitation at this latitude. In the Tihāma coastal plain the influence of summer rains ceases at al-Līṯ (Al Lith).

Summer precipitation is pronounced in the hilly Tihāma [Miḫwā, Maḫā'il (Mahayil), Sūq 'Aibān (Suq Ayban), Ǧabal Faifā' (Jabal Fayfa)] and on the escarpment of the 'Asīr highlands (Saudā'), where the monsoon winds release a large part of their water in the form of orographic precipitation.

East of the escarpment the influence of the summer rains declines. It is pronounced in the mountain region [Balğurašī, an-Nimāṣ (An Nimas), Tanūma (Tanoma), Abhā, Ḫamīs Mušaiṭ, Ẓahrān al-Ǧanūb (Zahran al Janub)] and drops to 6–13 mm (July–October) at the edge of the study area [Turaba (Turabah), Bīša, Nağrān (Najran)]. 'Afīf (Afif) and as-Sulayyil (As Sulayyil) have no summer precipitation.

Besides quantity and time of occurrence, the variability of precipitation is important to vegetation. This refers to the average deviation in percent from the mean annual precipitation measured over a number of years.

Meteorological data were obtained from the Ministry of Agriculture and Water, ar-Riyāḍ. The only stations considered were those for which complete information was available from at least seven years. Table 1 gives a survey of these stations, showing in each case the mean annual precipitation, the number of years of observation and the calculated variability of precipitation.

The relatively moist areas in the montane region (Balğurašī, an-Nimāṣ, Saudā') and on the south-western slope (Maḫā'il) exhibit the smallest deviations from the mean annual precipitation (variability 16–25 %). The largest deviations (variability over 60 %) occur at the stations of al-Līṯ and Makka, where the annual precipitation is about 100 mm, and at al-Madīna and as-Sulayyil, which have 30–40 mm.

Evaporation measurements, which are important in a number of ways for understanding vegetation conditions, will be discussed here as well. Measurements were carried out in the *Acacia-Commiphora* xeromorphic woodland of the Tihāma and in the *Juniperus* woodland of the montane region of 'Asīr. Observations and results are available from the spring (3–8 April, 1982) and summer (11–16 Aug., 1982).

The measurements were made with Piché evaporimeters (green filter-paper disks, 3 cm diameter). Peak evaporation levels of 3.4 ml/h were measured in the *Acacia-Commiphora* xeromorphic woodland; there were no significant differences between the April and August values. These values correspond to the rate of 3.6 ml/h reported by Walter (1960) from Port Sudan on the Sudanese side of the Red Sea on 15 Aug. 1934 (with hot desert wind, rates of up to 5.1 ml/h were recorded for short periods).

Station	Mean annual precipitation	Number of years of observation	Variability of precipitation
an-Nimāṣ	537 mm	8	18 %
Saudā'	559 mm	12	21 %
Maḫā'il	440 mm	8	23 %
Balǧurašī	456 mm	9	24 %
Ṣabyā	116 mm	10	27 %
as-Sail al-Kabīr	198 mm	9	36 %
al-Ḥawīya	233 mm	9	37 %
Tanūma	361 mm	10	37 %
al-Muḏailif	96 mm	7	40 %
Turaba	138 mm	8	41 %
Naǧrān	79 mm	8	44 %
Abū 'Arīš	142 mm	9	52 %
Ǧidda (Jeddah)	69 mm	10	56 %
Bīša	137 mm	9	58 %
al-Madīna	40 mm	10	64 %
Makka	114 mm	9	68 %
al-Līṯ	93 mm	9	72 %
as-Sulayyil	31 mm	7	145 %

Tab. 1 List of stations included for the calculation of variability of precipitation, grouped according to variabilities of 16–25, 26–40, 41–60, and over 60 %.

The daily evaporation rates in the *Acacia-Commiphora* xeromorphic woodland, measured at 34.6–40.5 ml/d in August, markedly exceeded those in April (21.5–25.0 ml/d) since evaporation was also comparatively high in the evening and night hours. Because of the high content of suspended particles in the air less heat is radiated in summer, with the result that night temperatures seldom dropped below 35° C during the period of observation.

The maximum evaporation values in the *Juniperus* woodland varied between 1.2 and 1.7 ml/h (data for August), depending on the density of the stands. Daily evaporation, recorded here at 2.4–4.7 ml/d (April) and 13.0–15.2 ml/d (August), was considerably less than that in the *Acacia-Commiphora* xeromorphic woodland. The differences between April and August are much greater here on account of the relatively low temperatures and the frequent reduction of sunlight by cloud cover in April.

Among the regional climatic factors that cannot be neglected are hail showers, which occasionally occur in the montane region above 2000 m. Besides mechanical damage (hailstones up to 1 cm in diameter have been observed) they can cause rapid temperature drops, which influence the vegetation particularly near ground level.
BARTH (1976) mentions snowfalls in the ʿAsīr highlands. For example, in the winter of 1949 Abhā had a closed snow cover of 10 cm for several days. Reports of snow in the Yemen Arab Republic are given in RATHJENS/WISSMANN (1934).

4. Zonation of vegetation

Whereas travellers in former times had only a few possibilities of traversing the ʿAsīr highlands from west to east, today there are a number of pass roads, making it possible to investigate the vegetation of the border escarpment at various places and to point out parallels. In the following, five profiles will be discussed. Fig. 1 shows their locations. They are based on information compiled from phytosociological samples and on detailed records of the terrain.
The first locations that will be discussed are the slopes and plateaus. Then the draining runnels and wadis will be examined. Special attention will be given to vegetation conditions at sites that today can still be regarded as undisturbed. The purpose is to describe the state of the 'natural landscape' that reflects the ecological conditions (climate, altitudinal zones, etc.). At the same time the aim is to create a basis for documenting the vegetation, for the expansion of the road network and the development of hitherto inaccessible areas can be expected to cause increasing damage and destruction at previously relatively undisturbed locations.

4.1 Profile 1: Tanūma (NE–SW) (Fig. 4)

Profile 1 stretches from the hilly Tihāma at about 500 m altitude to the mountainous region (2200 m) along the Tanūma pass road.
The hilly Tihāma and the escarpment up to 1600 m are inhabited by xeromorphic woodland of varying composition as shown in Fig. 4. In the lower altitudes species of the genus *Commiphora* are dominant besides *Acacia asak* and *A. tortilis*, whereas in the higher altitudes the xeromorphic woodland is mainly characterized by *Acacia asak* and *A. etbaica*.
Below 550 m *Acacia tortilis* is dominant with *Commiphora myrrha* and *C. opobalsamum* as co-dominant species (*Acacia tortilis-Commiphora* xeromorphic woodland). Above 550 m *Acacia tortilis* is replaced by *A. asak*

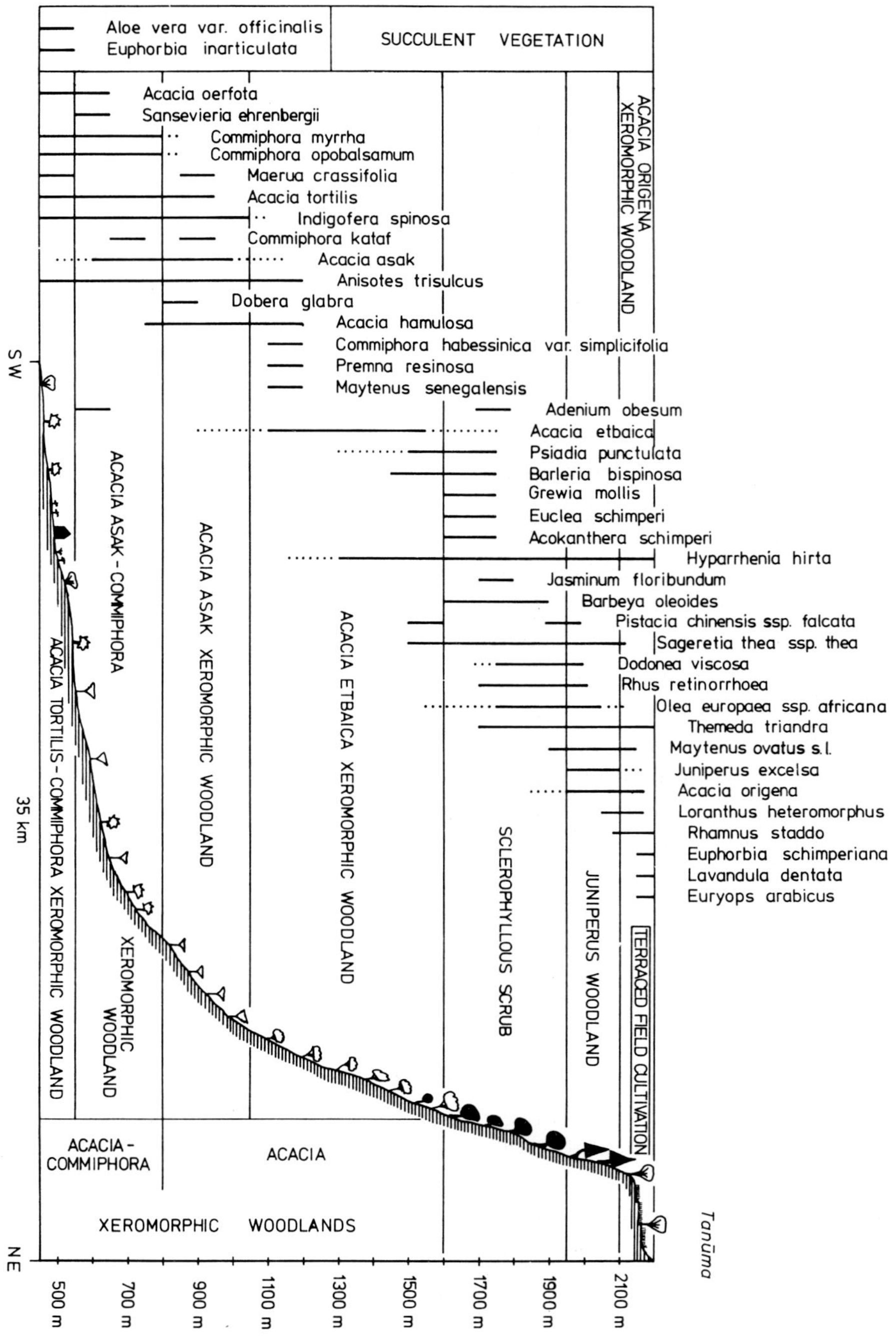

Fig. 4 Profile 1 across the south-western escarpment near Tanūma.

(*Acacia asak-Commiphora* xeromorphic woodland). Above 800 m *Commiphora myrrha* and *C. opobalsamum* withdraw; only *C. habessinica* var. *simplicifolia* (= *C. abyssinica* var. *simplicifolia*) and *C. kataf* are found sporadically. Thus, the vegetation is characterized at 800 to 1050 m by *Acacia asak* and at 1050 to 1600 m by *A. etbaica* xeromorphic woodland.

Further characteristic species of these zones include *Maerua crassifolia*, *Indigofera spinosa*, *Anisotes trisulcus*, *Dobera glabra*, *Acacia hamulosa*, *Premna resinosa*, *Maytenus senegalensis* and *Adenium obesum*.

In the vicinity of settlements of the lower altitudes the abundant presence of *Acacia oerfota (= A. nubica)* is conspicuous. This species is avoided by the cattle on account of its repugnant odour. Besides, succulent vegetation is found consisting of *Aloe vera* var. *officinalis* and *Euphorbia inarticulata*, which are often accompanied by *Cissus quadrangularis* and *C. rotundifolia*, less frequently by *Sansevieria ehrenbergii*. *Anisotes trisulcus* is met with at similar locations in high cover-abundance too.

In the transition to the inferior montane zone of the escarpment sclerophyllous scrub is found between 1600 and 1950 m (cf. Photo 2). The characteristic species comprise *Euclea schimperi (= E. kellau)*, *Acokanthera schimperi*, *Barbeya oleoides*, *Pistacia chinensis* ssp. *falcata*, *Sageretia thea* ssp. *thea (= Berchemia yemensis)*, *Dodonea viscosa*, *Rhus retinorrhoea* and *Olea europaea* ssp. *africana (= O. chrysophylla)*. Associated are *Grewia mollis*, *Jasminum floribundum* and as transition species to the *Acacia etbaica* xeromorphic woodland *Psiadia punctulata (= P. arabica)* and *Barleria bispinosa*. *Adenium obesum* reaches the highest stands here, but was seen only sporadically compared with Profile 3.

Moreover, grasses like *Hyparrhenia hirta* and *Themeda triandra* must be mentioned, which distinctly increase at about 1200 m. A striking feature is the wealth of annuals between 1600 and 1950 m, which gives this zone additional attractiveness in the 'rainy season'.

Among them are found *Alyssum desertorum*, *Ammi majus*, *Arenaria leptoclados*, *Biserrula pelecinus* ssp. *pelecinus*, *Bupleurum gerardii*, *Erodium subtrilobum*, *Fumaria parviflora*, *Gastridium phleoides*, *Geranium trilophum*, *Lamarckia aurea*, *Linum corymbulosum*, *Lophochloa pumila*, *Medicago laciniata*, *M. minima*, *M. polymorpha*, *Ononis reclinata*, *Plantago afra*, *Polycarpon tetraphyllum*, *Silene apetala*, *Torilis arvensis*, *Trifolium arvense*, *T. campestre*, *Valantia hispida* and *Vermifrux abyssinica* with variable cover-abundance values.

The montane zone between 1950 and 2200 m is occupied in the Tanūma region by a human-caused vegetation mosaic of woodland with *Juniperus excelsa* (= *J. procera* according to Kerfoot 1975, Kerfoot/Lavranos 1984) and xeromorphic woodland with *Acacia origena* (= *A. negrii* auct., *A. abyssinica* auct.; cf. remarks in chap. 5). On plateau areas terraced field cultivation is practised.

Further characteristic species of this zone include *Sageretia thea* ssp. *thea*, *Maytenus ovatus* s.l., *Rhamnus staddo*, *Euphorbia schimperiana*, *Lavandula dentata* and *Euryops arabicus*. *Loranthus heteromorphus* (incl. *L. regularis*) is a typical parasite of these altitudes and sponges in *Acacia origena*, less frequently on *Acacia gerrardii* or *Ziziphus spina-christi*.

4.2 Profile 2: Slope, hilly Tihāma, E al-Qunfuḏa (Fig. 5, Photo 1)

By means of the transect method 'relative importance values' were determined for the woody plants (phanerophytes) along an altitudinal gradient (cf. Mueller-Dombois/Ellenberg 1974) (Profile 2, Fig. 5). In the following the vegetational zonation of a slope in the hilly Tihāma will be described as an example.

The relative importance value is defined as the sum of relative density and relative dominance. To record density and dominance sample areas (10 m × 50 m) were marked out perpendicular to the gradient of the slope and estimates were made for each of them. Level areas are occupied mainly by *Acacia tortilis* and *Commiphora myrrha* (*Acacia tortilis-Commiphora myrrha* association). At the foot of the slope *Acacia hamulosa* and *Delonix elata* are found as well; *Acacia tortilis* recedes (*Commiphora myrrha-Acacia tortilis-Acacia hamulosa* association). The slope itself is occupied primarily by *Commiphora opobalsamum* and *Delonix elata* (*Commiphora opobalsamum-Delonix elata* association). *Acacia asak*, which is characteristic of this altitudinal zone, is conspicuously absent here. It cannot compete on the outcropping slaty rock and is replaced by *Acacia etbaica* in other comparable habitats.

Cadaba longifolia, *Maerua crassifolia*, *Euphorbia cuneata*, *Anisotes trisulcus* and *Grewia tenax* show no distinct zonation, but are nevertheless characteristic species of the *Acacia-Commiphora* xeromorphic woodland.

4.3 Profile 3: Ǧabal Saudā'–ad-Darb (NNE–SSW) (Fig. 6)

Profile 3 shows the conditions in the area of the south-west slope of Ǧabal Saudā' west of Abhā. On account of the relatively high precipitation and the absence of interference to date, the vegetation units have been preserved in a relatively natural state, in sharp contrast to the units along the road between Abhā and ad-Darb, which show marks of human interference especially at higher elevations. The frequent dwarf-shrubs, xeromorphic ferns and representatives of montane rock vegetation are also shown in profile 3.

In the zone at 250 to 400 m (*Acacia tortilis-Commiphora* xeromorphic woodland) *Acacia tortilis*, *Commiphora myrrha*, *C. opobalsamum* and *Euphorbia cuneata* dominate. *Commiphora kataf*, *C. habessinica* var. *simpli-*

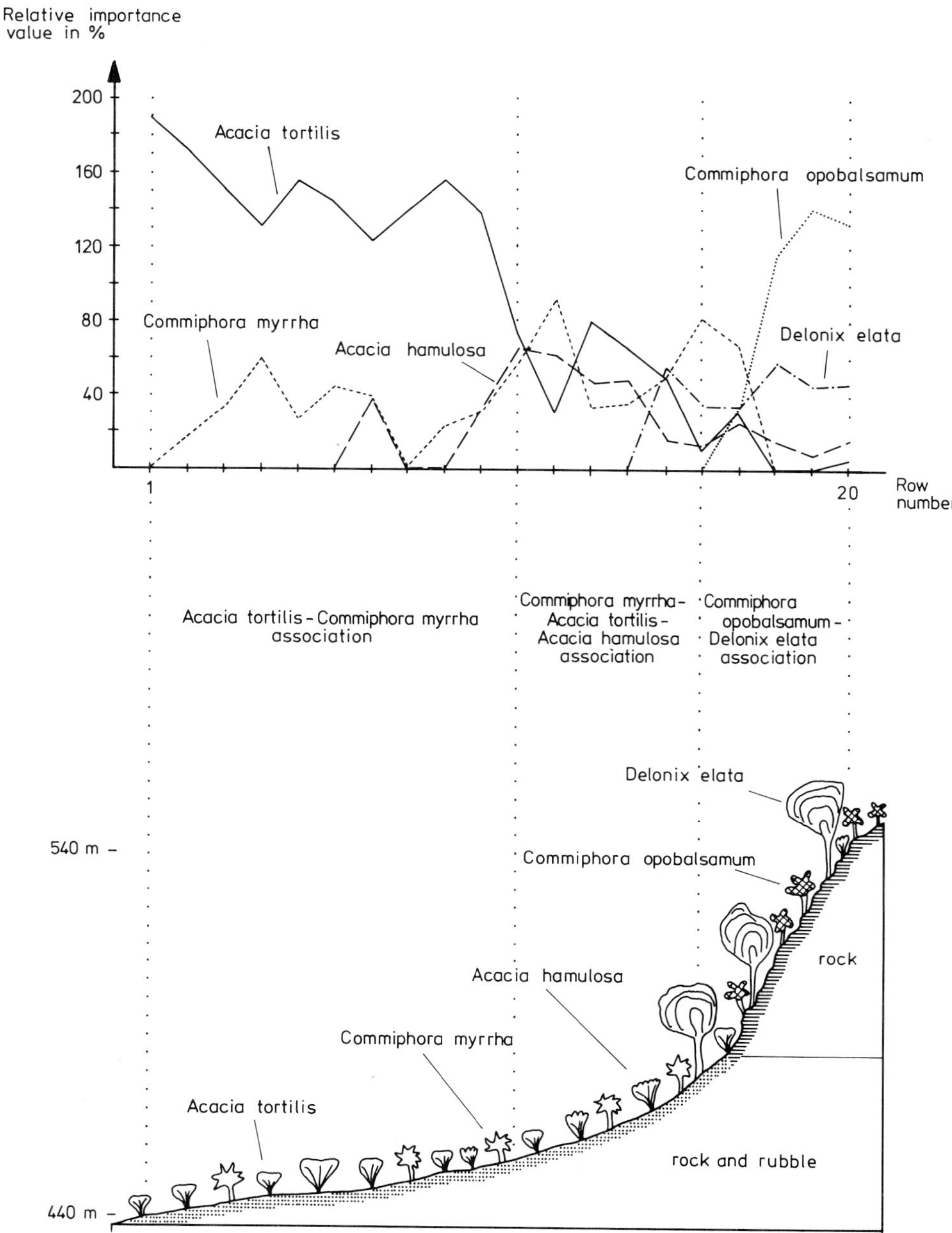

Fig. 5 Zonation of vegetation on a slope of the hilly Tihāma (Profile 2), 75 km E al-Qunfuḏa near Ǧidda – Maḫā'il road, eastern exposure, slaty rocks.

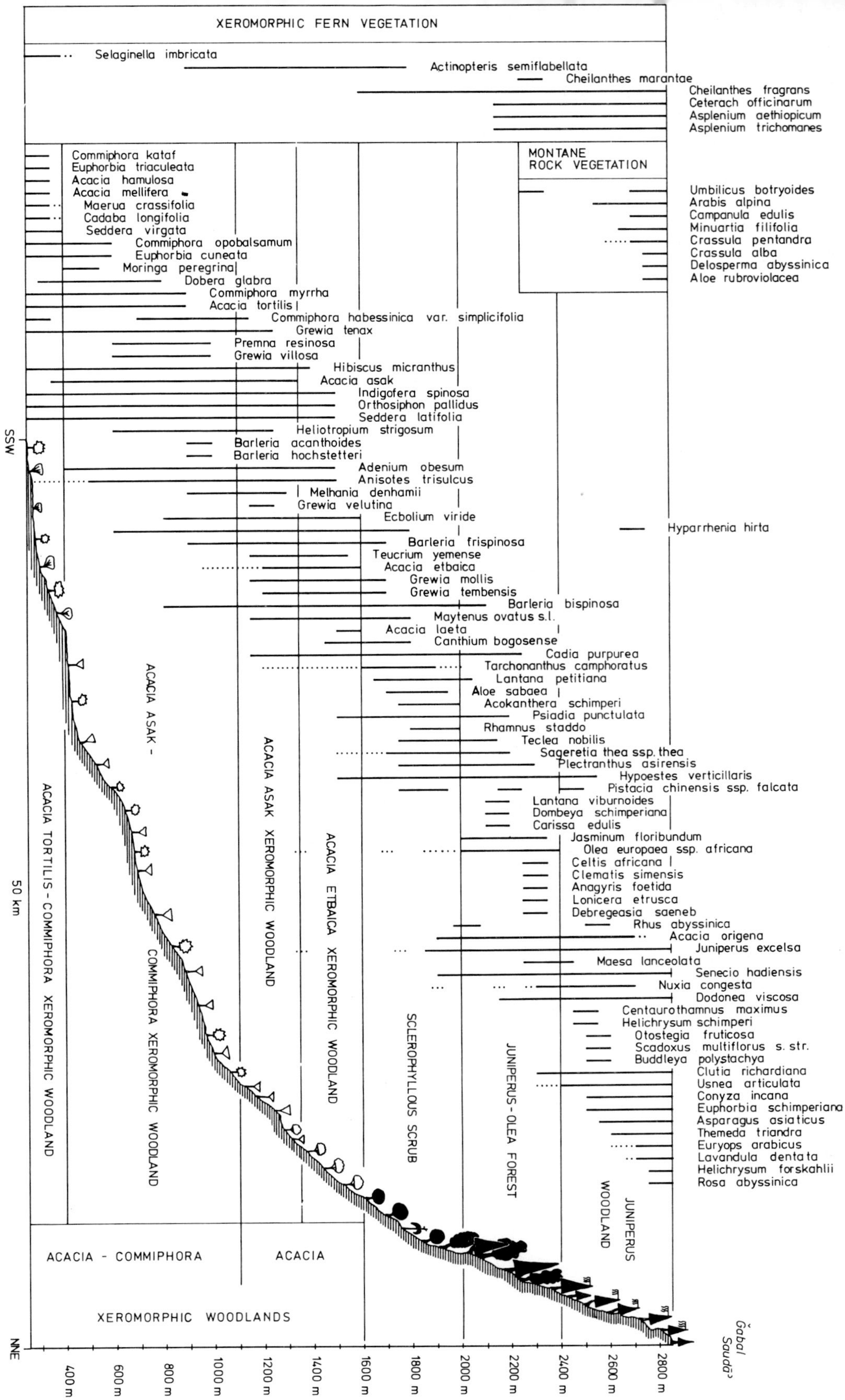

Fig. 6 Profile 3 across the south-western escarpment of Ğabal Saudā' (W Abhā) in the direction of ad-Darb.

cifolia, *Euphorbia triaculeata*, *Acacia hamulosa*, *A. mellifera*, *Maerua crassifolia*, *Cadaba longifolia* and *Dobera glabra* are represented with smaller cover-abundance.

Between 400 and 1100 m an *Acacia asak-Commiphora* xeromorphic woodland appears. Among the characteristic species are *Acacia asak*, *Commiphora myrrha*, *C. opobalsamum*, *C. habessinica* var. *simplicifolia*, *Euphorbia cuneata*, *Dobera glabra*, *Moringa peregrina*, *Premna resinosa*, *Grewia villosa* and *Acacia tortilis.*

Above 1100 m the vegetation is mainly characterized by *Acacia* species with *A. asak* at 1100 to 1350 m (*A. asak* xeromorphic woodland) and *A. etbaica* at 1350 to 1600 m (*A. etbaica* xeromorphic woodland). Associated with the *Acacia* xeromorphic woodland are *Grewia velutina*, *G. mollis*, *G. tembensis* and *Barleria trispinosa*, as well as *Barleria bispinosa*, *Maytenus ovatus* s.l., *Canthium bogosense* and *Cadia purpurea*, which merge already with the zone of sclerophyllous scrub.

Species with a large ecological amplitude, which are found in the total range of the xeromorphic woodlands, include *Grewia tenax*, *Adenium obesum*, *Anisotes trisulcus* and *Ecbolium viride* among the phanerophytes. Noteworthy is the wealth of dwarf-shrubs in the xeromorphic woodlands. *Seddera virgata*, *S. latifolia*, *Hibiscus micranthus*, *Indigofera spinosa*, *Orthosiphon pallidus*, *Heliotropium strigosum*, *Barleria acanthoides* and *B. hochstetteri* occur with variable cover-abundance and different altitudinal distribution.

The xeromorphic woodlands are followed by sclerophyllous scrub, which characterizes the zone between 1600 and 2000 m. This zone includes, besides *Acokanthera schimperi*, *Teclea nobilis* and *Pistacia chinensis* ssp. *falcata*, the characteristic tree-like Composite *Tarchonanthus camphoratus.* Some species like *Cadia purpurea*, *Lantana petitiana*, *Psiadia punctulata*, *Rhamnus staddo*, *Sageretia thea* ssp. *thea*, *Plectranthus asirensis* and *Hypoestes forskalei* (= *H. verticillaris* auct.) overlap to the next zone. *Aloe sabaea*, with up to 2-m-high trunks, is a striking species among the succulents in the sclerophyllous scrub, but was found in other localities at lower altitudes (1000–1500 m).

The sclerophyllous zone is followed by a *Juniperus-Olea* forest with *Juniperus excelsa* and *Olea europaea* ssp. *africana* (2000–2400 m). This forest type observed on Ǧabal Saudā' is one of the most impressive community types of the study area and was found in this composition here only.

The occurrence of *Teclea nobilis*, *Dombeya schimperiana* (= *D. hirsuta*), *Celtis africana* (= *C. kraussiana*), *Debregeasia saeneb* (= *D. salicifolia*), *Rhus abyssinica*, *Maesa lanceolata*, *Nuxia congesta* and peripherally also *Buddleja polystachya* in the *Juniperus-Olea* forest is evidence for floristic relations to the *Juniperus-Podocarpus* forests of Ethiopia (GILLETT 1941, KNAPP 1968, 1973).

Further species are *Lantana viburnoides*, *Carissa edulis*, *Anagyris foetida* and *Acacia origena*; the lianas include *Jasminum floribundum*, *Clematis simensis* and *Lonicera etrusca*.

Above 2400 m up to the mountain peaks the vegetation is characterized by *Juniperus* woodland (Photo 3). *Acacia origena* is associated with *Juniperus excelsa* up to 2700 m.

As a result of cloud formation and frequent passage of fog over the peaks, epilithic and epiphytic communities develop. Two sites were investigated by Kürschner (1984a, b). The most typical representative is the beard lichen *Usnea articulata*, which occurs from about 2400 m up to the summits at locations suitably exposed to the moving fog (SW exposure).

The undergrowth in the *Juniperus* woodland is made up of evergreen shrubs like *Dodonea viscosa*, *Euryops arabicus* and *Lavandula dentata* as well as *Clutia richardiana*, *Conyza incana* and *Euphorbia schimperiana*. *Rosa abyssinica* prefers moister places. *Asparagus asiaticus* climbs in the branches of *Juniperus excelsa* and *Acacia origena*.

More or less barren rock formations are inhabited by rock vegetation like that of the higher mountain regions of Ethiopia and Somalia. Among the occurring species are *Umbilicus botryoides*, *Arabis alpina*, *Campanula edulis*, *Minuartia filifolia*, *Crassula pentandra*, *C. alba*, *Delosperma abyssinica* and *Aloe rubroviolacea*.

Xeromorphic fern vegetation is found in crevices and under boulders. This community type includes *Asplenium aethiopicum*, *A. trichomanes*, *Ceterach officinarum* and *Cheilanthes maranthae* in the high altitudes, *Cheilanthes fragrans* as well as *C. catanensis*, *C. coriacea* and *Onychium divaricatum (= O. melanolepis)* in the middle altitudes and *Selaginella imbricata* and *Actinopteris semiflabellata* in the range of the xeromorphic woodlands. *Pleopeltis macrocarpa* has higher demands on the water supply and was found only once within the *Juniperus-Olea* forest.

4.4 Profile 4: Ǧabal Saudā' (WSW–ENE) (Fig. 7)

Whereas the previous profiles showed vegetation conditions on the ʿAsīr escarpment facing the Red Sea and in the hilly Tihāma, in the following the eastern slopes will be dealt with.

The *Juniperus* woodland of the summit region is followed on Ǧabal Saudā' between 2650 and 2500 m and outside the terraced field cultivation areas by community types which are dominated either by *Juniperus excelsa* or *Acacia origena*. These merge via *Acacia origena* xeromorphic woodland into evergreen open xeromorphic scrub (2450–2300 m). Characteristic representatives, which were partly already mentioned for the undergrowth of the *Juniperus* woodland, include *Euphorbia schimperiana*, *Lavandula den-*

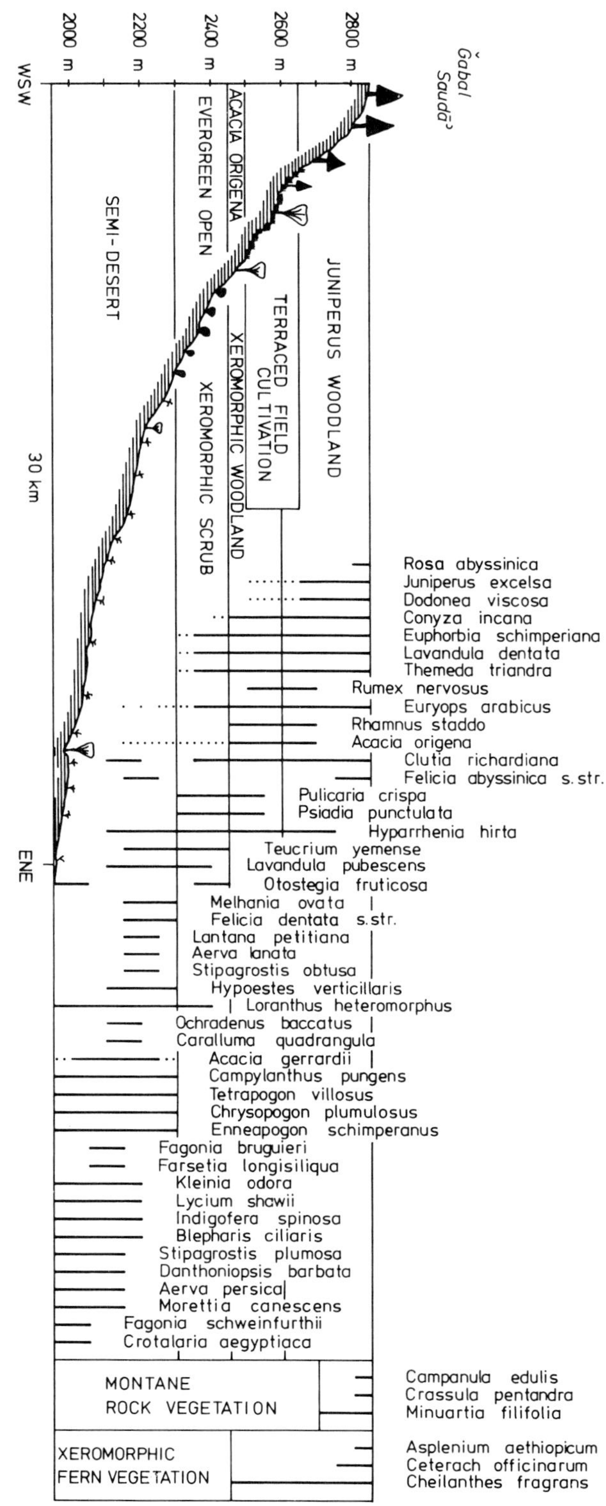

Fig. 7 Profile 4 across the north-eastern slopes of Ǧabal Saudā', N Abhā.

tata, Euryops arabicus, Clutia richardiana, Pulicaria crispa and *Psiadia punctulata* in the evergreen open xeromorphic scrub, and *Themeda triandra* and *Hyparrhenia hirta* among the grasses.

Below 2300 m the vegetation gets more and more sparse and there appear semi-desert community types with scattered dwarf-shrubs, grasses, etc. (Photo 4). Among the dwarf-shrubs are *Melhania ovata, Felicia dentata* ssp. *dentata, Aerva lanata, A. persica, Campylanthus pungens, Fagonia bruguieri, F. schweinfurthii, Farsetia longisiliqua, Indigofera spinosa, Morettia canescens* and *Crotalaria aegyptiaca*, which have variable cover-abundance values; among the grasses are *Stipagrostis obtusa, S. plumosa, Tetrapogon villosus, Chrysopogon plumulosus* (= *C. aucheri* var. *quinqueplumis*), *Enneapogon schimperanus* and *Danthoniopsis barbata.*

In runnels and wadis there occur the shrubs and trees *Lycium shawii, Acacia origena* (2700–2150 m) and *A. gerrardii* (below 2300 m); the latter are often studded with *Loranthus heteromorphus.* In the vicinity of villages succulent vegetation with *Kleinia odora* is found in this zone. The montane rock vegetation and xeromorphic fern vegetation overlap only a little and in impoverished composition on the slopes east of the summit region.

4.5 Profile 5: al-ʿUlayya – Bīša (SW – NE) (Fig. 8)

The conditions of vegetation in the al-ʿUlayya region are shown in Fig. 8. The mountains reach about 2200 m and the escarpment to the Red Sea is comparatively moderate so that rain shadow effects along the eastern slopes are not so marked as on Ǧabal Saudā' (Profile 4, Fig. 7). The boundaries between the vegetation units are less clearly pronounced.

The high altitudes at 2200–2100 m are occupied by a *Juniperus* woodland with *Juniperus excelsa, Euryops arabicus, Euphorbia schimperiana* and *Clutia richardiana.*

Below 2100 m *Acacia gerrardii* becomes more characteristic of the vegetation. Whereas in the beginning *Juniperus-Acacia gerrardii* woodland with *Clutia richardiana, Sageretia thea* ssp. *thea* and *Dodonea viscosa* is found, this community type is replaced by *Olea-Acacia gerrardii* woodland below 2000 m. Characteristic species of the latter are *Acacia gerrardii, Olea europaea* ssp. *africana, Barbeya oleoides, Pistacia chinensis* ssp. *falcata* and *Jasminum floribundum.*

In habitats with blocky weathered granites the above-mentioned species are joined in the *Olea-Acacia gerrardii* zone by *Rhus retinorrhoea, Euclea schimperi, Carissa edulis* and *Ficus salicifolia.* Cover-abundance of *Olea europaea* ssp. *africana, Dodonea viscosa, Hypoestes forskalei* and *Psiadia punctulata* increases so that local sclerophyllous scrub is present (Photo 2).

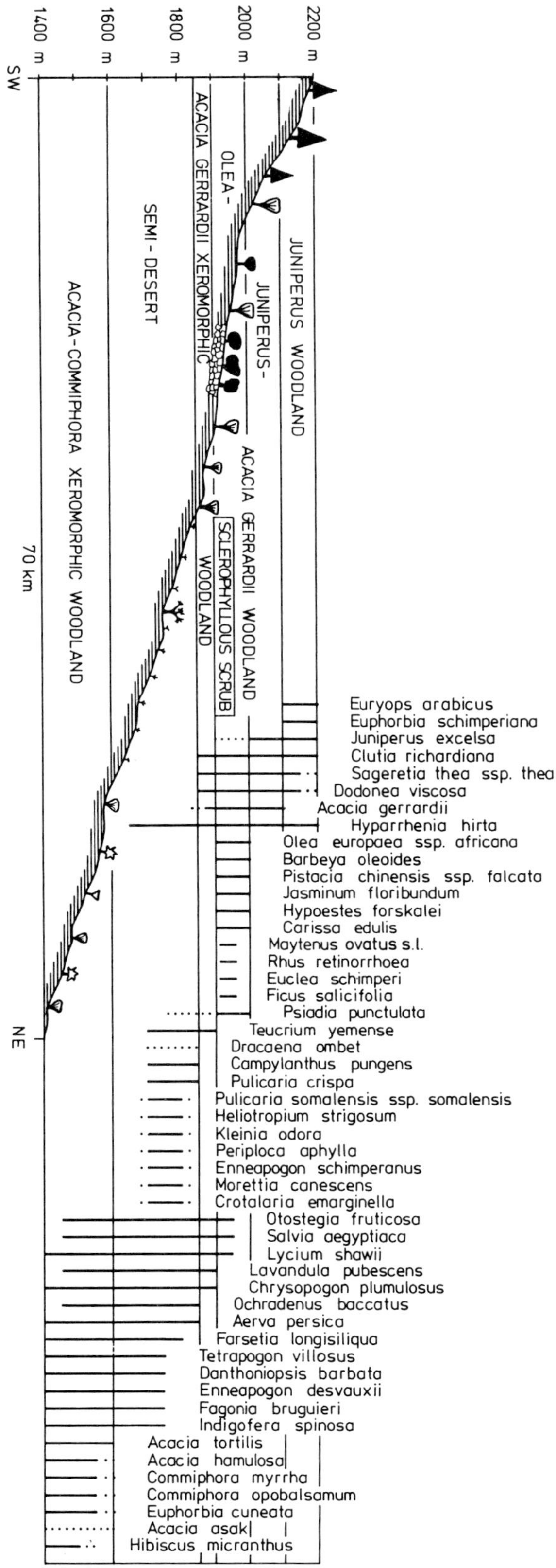

Fig. 8 Profile 5 across the north-eastern slopes near al-ʿUlayya in the direction of Bīša.

With increasing distance from the watershed only *Acacia gerrardii* xeromorphic woodland is found, which is replaced below 1850 m by semi-desert characterized by scattered dwarf-shrubs etc. The dwarf-shrubs include *Teucrium yemense, Campylanthus pungens, Pulicaria crispa, P. somalensis* ssp. *somalensis, Heliotropium strigosum, Morettia canescens, Crotalaria emarginella, Salvia aegyptiaca, Lavandula pubescens, Aerva persica, Farsetia longisiliqua, Fagonia bruguieri* and *Indigofera spinosa* with varying cover-abundance.

In addition *Kleinia odora, Periploca aphylla, Otostegia fruticosa, Lycium shawii* and *Ochradenus baccatus* occur; among the grasses *Enneapogon schimperanus, Chrysopogon plumulosus, Tetrapogon villosus, Danthoniopsis barbata* and *Enneapogon desvauxii (= E. brachystachyus)* must be mentioned.

A particularly conspicuous species is *Dracaena ombet*, found between 1850 and 1700 m. It is not confined to the eastern slopes but occurs in the southern part of the study area as well, where it grows in a number of places on the south-western slope of the border range. This is of particular interest since in Africa *Dracaena ombet* must be considered a highly endangered species (IUCN 1978).

Below 1600 m *Acacia-Commiphora* xeromorphic woodland with *Acacia asak* (sporadically), *A. hamulosa, A. tortilis, Commiphora myrrha, C. opobalsamum* and *Euphorbia cuneata* occurs.

4.6 Draining runnels and wadis (Fig. 9)

The vegetation conditions in the runnels and wadis will be described here in general terms. For the sake of completeness the Tihāma coastal plain is included. Distribution data for the dominant species are compiled in Fig. 10.

Owing to the relatively high amounts of precipitation in the upper parts of the ʿAsīr highlands (400–600 mm/year) there is considerable drainage, and this is reflected in the luxuriance of vegetation along the drainage systems. The following zones, related mainly to orographic factors, can be distinguished. The south-western slopes and the montane region will be dealt with first.

Between 2800 and 2000 m *Buddleja polystachya, Erica arborea, Hypericum hircinum, H. revolutum (= H. lanceolatum), Maesa lanceolata, Maytenus undatus, Myrica salicifolia, Nuxia congesta, Salix* sp. and *Rosa abyssinica* are found around pools and in the draining runnels, which are mostly carved in the rocky subsoil.

Further down *Breonadia salicina (= Adina microcephala), Combretum molle* (up to 200 m), *Cordia ovalis, Ficus lutea (= F. ingens), F. sycomorus, Nuxia*

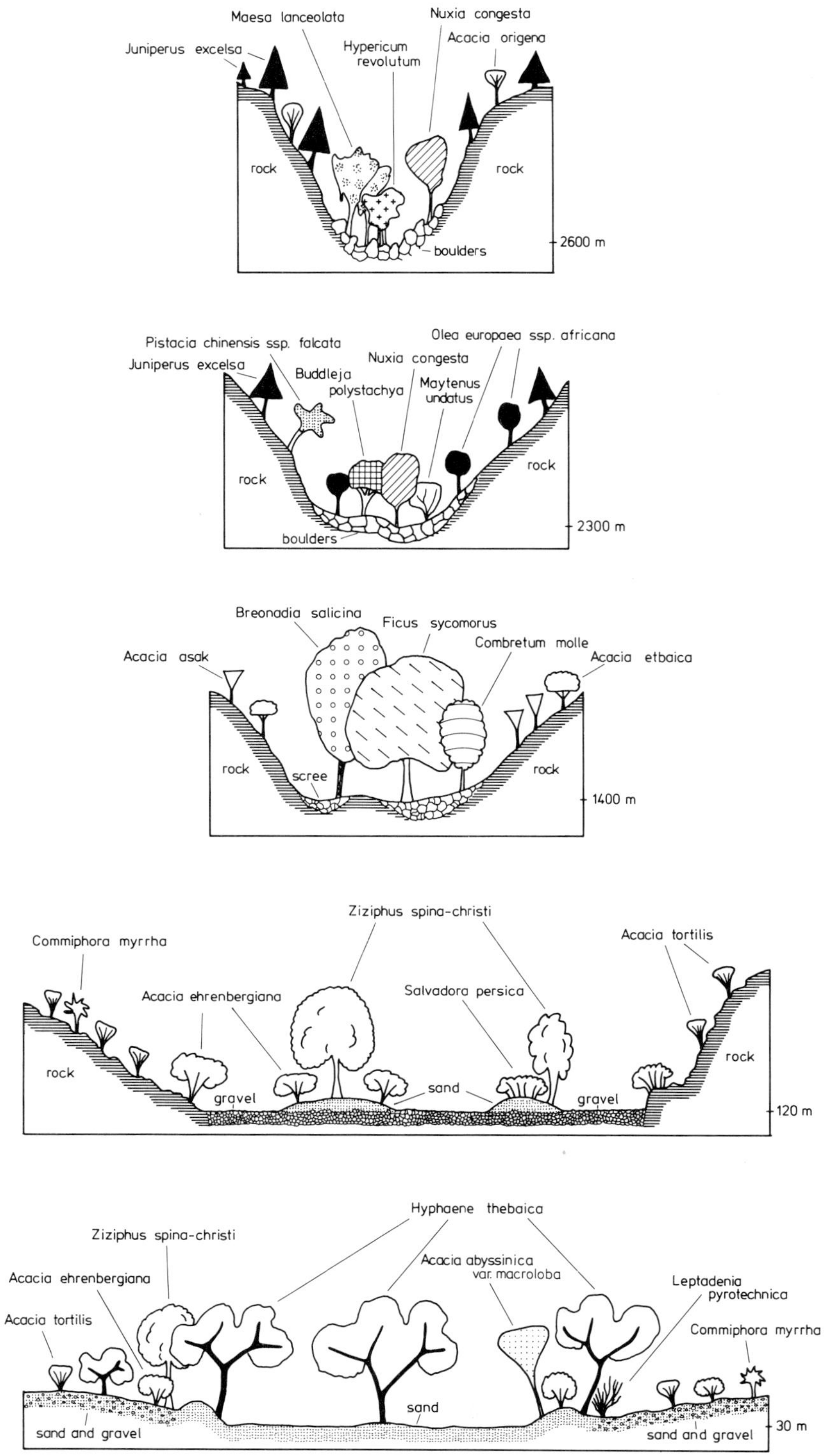

Fig. 9 Profiles across draining runnels and wadis from different altitudinal zones, only the dominant species considered.

oppositifolia, *Phoenix reclinata* and *Ziziphus spina-christi* (up to sea-level) occur with increasing flow and transition of the runnels to wadis at altitudes between 2000 and 1000 m. Wadi species with a wide range of distribution which are well represented in this zone include *Ficus palmata* (2700–1000 m) and *F. salicifolia* (2200–300 m).

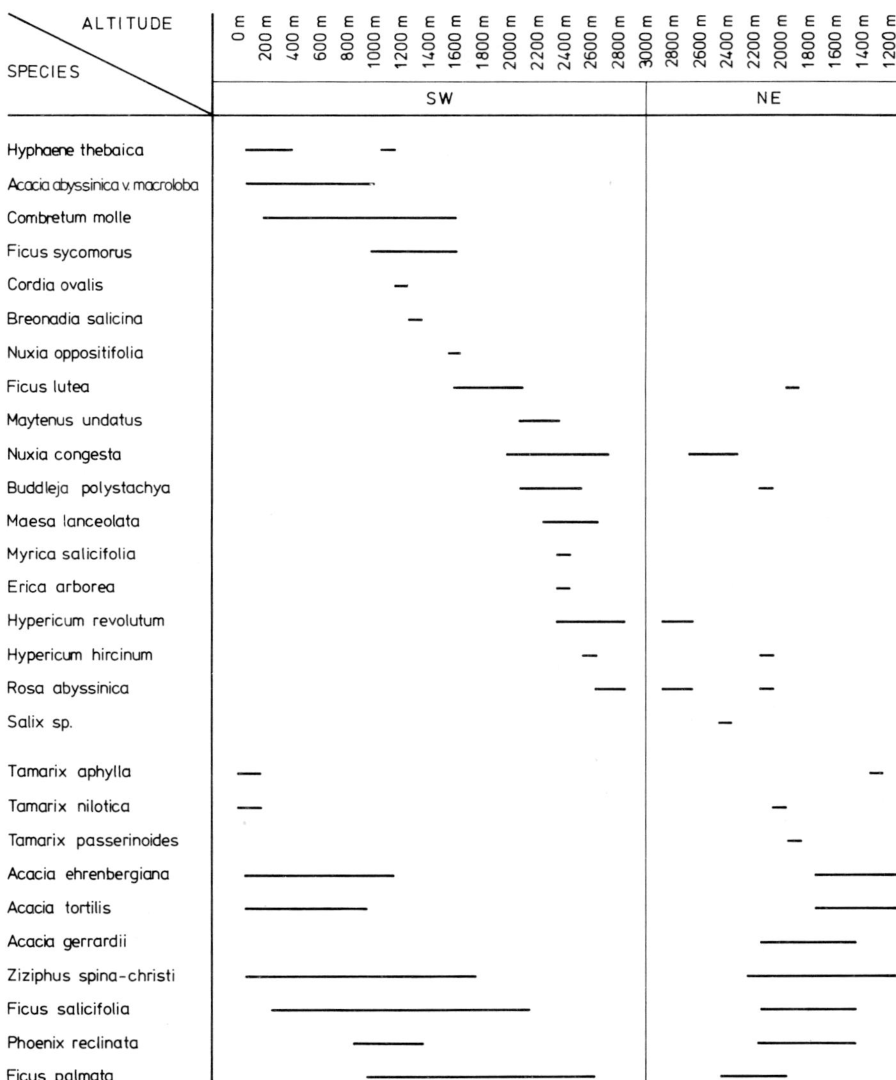

Fig. 10 Altitudinal distribution of the dominant species of draining runnels and wadis. Further unidentified *Tamarix* spp. between 400 and 600 m (SW) and between 1300 and 1250 m (NE).

With the transition to the hilly Tihāma and with decreasing relief intensity the flow rate is reduced, the wadis become shallow and there is increased deposition of fine material. Beginning at 1000 m there are species that are limited strictly to wadis, such as *Acacia abyssinica* var. *macroloba*, *Combretum molle* and *Ziziphus spina-christi*, as well as others that show a certain affinity to runnels or wadi peripheries, such as *Commiphora kataf*, *Dobera glabra*, *Moringa peregrina* and *Salvadora persica*.
Broad wadis and expansive valleys connected with them are occupied by xeromorphic woodland consisting mainly of *Acacia ehrenbergiana* and *A. tortilis*. *Acacia ehrenbergiana* prefers a subsoil rich in fine earth, while *A. tortilis* prefers a rocky-gravelly subsoil.
In the Tihāma coastal plain with the fan-shaped sandy wadi systems the characteristic species are *Hyphaene thebaica*, *Acacia abyssinica* var. *macroloba*, *Cadaba rotundifolia*, *Salvadora persica*, *Tamarix aphylla* and *T. nilotica*.
The drainage systems of the eastern slopes are characterized by a more gentle gradient and lower runoff rates. The dominating species are *Acacia ehrenbergiana*, *A. gerrardii*, *A. tortilis*, *Ficus palmata*, *F. salicifolia*, *Phoenix reclinata*, *Salvadora persica* and *Ziziphus spina-christi*. Where the wadi systems have a greater water flow and greater deposition of fine earth, *Tamarix* species are again able to establish themselves.
In the vicinity of villages and roads the wadi vegetation is often degraded and the stands are occupied in the lower and middle altitudes by *Calotropis procera* and/or *Rhazya stricta*.

5. Conclusion, phytosociological and plant-geographical considerations, discussion

In the following a summary is given of vegetation conditions on the basis of the profiles described. Phytosociological classification proposals for the Afro-Arabian region (KNAPP 1965, 1968, ZOHARY 1973) are examined and compared, and considerations related to plant geography are discussed.
The hilly Tihāma and the upper and middle parts of the south-western ʿAsīr slopes, along with some middle-altitude areas on the eastern slopes, are occupied by xeromorphic woodland. The vegetation here is dominated by species of the genera *Acacia* and *Commiphora*, as well as *Euphorbia cuneata*. There are also *Cadaba* and *Maerua* species. In the southern part of the study area, where there is a greater amount of summer rain (cf. Fig. 3), *Boscia angustifolia* and *Dobera glabra* are more frequent.
These units were classified by ZOHARY (1973) with the Acacietea sudano-arabica Zoh. 73; those of the eastern slopes were grouped with the Acacie-

tea tortilis subsudanica Zoh. 73 on account of their high percentage of Saharo-Arabian species.

According to KNAPP (1965, 1968, 1973) the xeromorphic woodlands in south-western Saudi Arabia belong to the Acacietea tortilis Kn. 68 and the Boscio-Commiphoretea Kn. 65. Here we shall follow this division scheme because its greater amount of detail facilitates classification (information on distribution is given in brackets according to KNAPP l. c.).

According to the present relevés the vegetation units of the lower altitudes with *Commiphora opobalsamum*, *C. myrrha*, *C. habessinica* var. *simplicifolia*, *C. quadricincta*, *Delonix elata*, *Cadaba longifolia*, *C. glandulosa* and *Maerua oblongifolia* are closely connected with the Commiphoretalia abyssinicae Kn. 68 (Sudan, N Ethiopia incl. Eritrea), though the presence of *Acacia mellifera*, *A. tortilis* and *A. hamulosa* shows relations to the Acacietalia tortilis Kn. 68 (Sudan, Ethiopia incl. Eritrea and N Somalia). The southern part of the hilly Tihāma in the border-area to the Yemen Arab Republic is inhabited by pure Commiphoretalia abyssinicae Kn. 68 vegetation units (W. Frey, pers. comm. 1984).

The xeromorphic woodland of the extensive wadis and valleys with *Acacia ehrenbergiana* and *A. tortilis* ranks amongst the Acacietalia tortilis Kn. 68. The vegetation of the corresponding wadi systems of the Tihāma coastal plain with *Hyphaene thebaica* merges into the Hyphaenetalia thebaicae Kn. 68.

The vegetation units at about 800 to 1600 m with *Acacia asak* and *A. etbaica* as characteristic species belong to the Acacietalia etbaicae Kn. 68 (Somalia, Ethiopia incl. Eritrea and E Sudan in middle altitudes).

With respect to plant geography the regions occupied by the above-described units belong to the 'Sudano-Zambezian region' (WICKENS 1976); close relations can be shown to exist between the 'Sahelian', 'Afrooriental' and 'South Arabian domain'. ZOHARY (1973) expressed this by distinguishing an 'Eritreo-Arabian province' within his 'Sudanian region', which corresponds to a large degree to the 'Sudano-Zambezian region' concerning the Afro-Arabian area. NEWTON (1980) proposed the term 'Mandeb circle' for these relations between NE Africa and SW Arabia.

The submontane sclerophyllous scrub is attached to the Pistacio-Eucleetalia schimperi Kn. 68 (N and E Ethiopia, N Somalia and E Sudan). The Hyperico-Rhamnetalia Kn. 68, which is represented in the study area around pools and at draining runnels between 2000 and 2800 m, belongs to the same class (Hyperico-Rhamnetea Kn. 68).

The vegetation units characterized by *Juniperus excelsa* belong to the Oleo-Juniperion aethiopicum Kn. 68 (Oleo-Juniperetalia procerae Kn. 68, Rapaneo-Macarangetea Kn. 65). The formations richest in precipitation on Ǧabal Saudā', which include *Dombeya schimperiana*, *Maesa lanceolata*, *Nuxia congesta* and other species (cf. remarks in chap. 4.3), already show resemblances to the Dombeyo-Podocarpion Kn. 68 (Oleo-Podocarpetalia Kn. 65, Rapaneo-Macarangetea Kn. 65).

Zohary (1973) proposed in his 'Geobotanical outline map of the Middle East' a classification in Oletea africanae, Juniperetea procerae and Podocarpetea for the montane region, but no details are given of species composition.

Vegetation units analogous to the xeromorphic woodland characterized by *Acacia origena* are known at high altitudes in north-eastern Africa, where they belong to the Acacietalia abyssinicae Kn. 68. *Acacia origena* can be regarded as a vicariad of *A. abyssinica*, and according to studies by Hunde (1982) the former is "no doubt closely related to both *A. abyssinica* and *A. negrii*". In the literature it has therefore been given the names of these two species (Hepper/Wood 1979, Kürschner 1984a as *'A. negrii'*; Schwartz 1939, Wissmann 1972 as *'A. abyssinica'*). The proportion of *Acacia origena* xeromorphic woodland has multiplied as a result of grazing and selective wood-gathering.

Plant-geographically, the units at high altitudes in 'Asīr belong to the 'Afromontane region' (White 1978). The species in this 'archipelago-like' floristic region are distributed over several mountain systems. In the study area they include *Hypericum revolutum*, *Maesa lanceolata*, *Myrica salicifolia* and *Nuxia congesta*. Transgressors from other floristic regions are *Euclea schimperi*, *Juniperus excelsa*, *Olea europaea* ssp. *africana*, *Celtis africana* and *Teclea nobilis*. The close relations between south-west Arabia and north-east Africa are demonstrated by the species *Barbeya oleoides*, *Acacia origena*, *Rosa abyssinica*, *Rhus abyssinica*, *R. retinorrhoea* and *Euryops arabicus* (with an exclave in the mountains of Oman), which are largely restricted to the 'Ethiopian system' of the 'Afromontane region'. Others such as *Acokanthera schimperi* and *Pistacia chinensis* ssp. *falcata* extend only slightly beyond this area and belong to the 'Eritreo-Arabian province' according to the classification of Zohary (1973).

At this point a look will be taken at the vegetation units that adjoin the montane forest zone in north-eastern Africa. Although the orographic climax community in the montane region of 'Asīr is made up of *Juniperus excelsa*, there are noteworthy relations to the '*Erica* thickets formation' described by Gillet (1941) for Ethiopia and the units called 'Altimontane scrub' by Pichi-Sermolli (1957) respectively. This is indicated by the occurrence of *Erica arborea*, *Hypericum revolutum*, *Rosa abyssinica*, *Myrica salicifolia*, *Myrsine africana*, *Clutia richardiana*, *Helichrysum schimperi*, *H. forskahlii* (= *H. fruticosum*), *Arabis alpina* and *Scabiosa columbaria*.

However, the orographic conditions and the humidity in the Saudi Arabian border mountains do not suffice for the formation of a separate vegetation unit analogous to that in Africa consisting mainly of *Erica arborea*.

Hepper/Wood (1979) describe 'large communities of *Erica arborea* on Jebel Melhan and Jebel Raymah' in the Yemen Arab Republic 'that Pichi-Sermolli would call altimontane scrub in Ethiopia'.

6. Summary

Beginning with a description of the climatic conditions, which are the decisive factor determining vegetational pattern in the hilly and mountainous regions of the Tihāma and ʿAsīr (SW Saudi Arabia), the zonation of vegetation units is analysed by means of five sample profiles of slope and plateau locations.
Acacia tortilis-Commiphora, *A. asak-Commiphora*, *A. asak* and *A. etbaica* xeromorphic woodland are characteristic of the lower and middle altitudes of the escarpment and the middle altitudes of the eastern slopes respectively. Typical communities in the montane region are sclerophyllous scrub, *Juniperus-Olea* forest, *Juniperus* woodland, *Juniperus-Acacia gerrardii* and *Olea-A. gerrardii* woodland, *A. gerrardii* and *A. origena* xeromorphic woodland, evergreen open xeromorphic scrub and semi-desert. Succulent vegetation, xeromorphic fern vegetation and montane rock vegetation are also found. The distribution of these communities is closely correlated with orographic, climatic, human and edaphic factors. The vegetation of the wadi systems is treated in summary form. Finally, a grouping according to the existing phytosociological classification schemes is carried out and some comments are made regarding floristic relations.

7. References

Abulfatih, H. A., Vegetation of higher elevations of Asir, Saudi Arabia, Proc. Saudi Biol. Soc. *3*, 1979, 139–148.

Abulfatih, H. A., Plant ecology of Dalaghan National Park, Asir province, Saudi Arabia, Proc. Saudi Biol. Soc. *5*, 1981 a, 131–141.

Abulfatih, H. A., Wild plants of Abha and its surroundings, Proc. Saudi Biol. Soc. *5*, 1981 b, 143–159.

Abulfatih, H. A., Elevationally restricted floral elements of the Asir mountains, Saudi Arabia, Journ. Arid Environm. *7*, 1984, 35–41.

Baierle, H. U. / El-Sheikh, A. M. / Frey, W. / König, P., / Kürschner, H., Central Saudi-Arabia. Vegetation, Map A VI 9 Tübinger Atlas Vorderer Orient, 1985, Wiesbaden.

Barth, K. H., Landesnatur, In: Blume, H. (Ed.), Saudi-Arabien – Natur, Geschichte, Mensch und Wirtschaft, 1976, 23–88, Tübingen & Basel.

Batanouny, K. H., Vegetation along the Jeddah-Mecca road: pattern and process as affected by human impact, Journ. Arid Environm. *2*, 1979, 21–30, 5 pls.

Batanouny, K. H. / Baeshin, N. A., Studies on the flora of Arabia. I. The Jeddah-Mecca road, Saudi Arabia, Taeckholmia *9*, 1978, 67–81.

Batanouny, K. H. / Baeshin, N. A., Plant communities along the Medina-Badr road across the Hijaz mountains, Saudi Arabia, Vegetatio *53*, 1983, 33–43.

Brooks, W. H. / Mandil, K. S. D., Vegetation dynamics in the Asir woodlands of south-western Saudi Arabia, Journ. Arid Environm. *6*, 1983, 357–362.

Chapman, R. W., 1.1 Geology, 1.2 Geomorphology, In: Al-Sayari, S. S. / Zötl, J. G. (Ed.), Quaternary period in Saudi Arabia. 1: Sedimentological, hydrogeological, hydrochemical, geomorphological, and climatological investigations in central and eastern Saudi Arabia, 1978, 4–30, Wien & New York.

Flohn, H., Klimaprobleme am Roten Meer, Erdkunde *19*, 1965, 179–191.

Geological Map of the Kingdom of Saudi Arabia 1 : 500 000, I-210 A Southern Hijaz, 1963, I-216 A Tihamat ash Sham, 1958, GM-217 A Asir, 1979 (Reprint of I-217 A, 1959), U. S. Geological Survey, Washington.

Gillett, J. B., The plant formations of western British Somaliland and the Harar province of Abyssinia, Kew Bull. 1941, 37–75.

Hepper, F. N. / Wood, J. R. I., Were there forests in the Yemen?, Proc. Seminar Arabian Studies *9*, 1979, 65–71.

Hunde, A., Two new species of *Acacia* (Leguminosae-Mimosoideae) from Ethiopia and Yemen, Nord. Journ. Bot. *2*, 1982, 337–342.

IUCN (Ed.), The IUCN plant red data book, compiled by G. Lucas & H. Synge, Morges, 1978, Switzerland.

Kassas, M., The mist oasis of Erkwit, Sudan, Journ. Ecol. *44*, 1956, 180–194.

Kerfoot, O., Origin and speciation of the Cupressaceae in Sub-Saharan Africa, Boissiera *24*, 1975, 145–150.

Kerfoot, O. / Lavranos, J. J., Studies in the flora of Arabia X: *Juniperus phoenicea* L. and *J. excelsa* M. Bieb., Notes RBG Edinb. *41*, 1984, 483–489.

Khattab, A. / El-Hadidi, M. N., Results of a botanic expedition to Arabia in 1944–45, Publ. Cairo Univ. Herb. *4*, 1971, 95 pp.

Knapp, R., Pflanzengesellschaften und höhere Vegetationseinheiten von Ceylon und Teilen von Ost- und Central-Afrika, Geob. Mitt. *33*, 1965, 1–31.

Knapp, R., Höhere Vegetationseinheiten von Äthiopien, Somalia, Natal, Transvaal, Kapland und einigen Nachbargebieten, Geob. Mitt. *56*, 1968, 1–36.

Knapp, R., Die Vegetation von Afrika, 1973, Stuttgart.

König, P., Vegetation und Flora im südwestlichen Saudi-Arabien (Asir, Tihama), Dissertation Freie Universität Berlin, in prep..

Kürschner, H., Epiphytic communities of the Asir mountains (SW Saudi Arabia). Studies in Arabian bryophytes 2, Nova Hedwigia *39*, 1984 a, 177–199.

Kürschner, H., An epilithic bryophyte community in the Asir mountains (SW Saudi Arabia). Studies in Arabian bryophytes 4, Nova Hedwigia *40*, 1984 b, 423–434.

Mandaville, J. P., A contribution to the flora of Asir, southwestern Saudi Arabia, Field Research Publications, Coconut Grove, Miami, 1973, iv + 13 pp.

Mueller-Dombois, D. / Ellenberg, H., Aims and methods of vegetation ecology, 1974, New York, Chichester, Brisbane & Toronto.

Newton, L. E., Phytogeographical associations of the succulent plant flora of south-west Arabia and the Horn of Africa, Nat. Cact. Succ. Journ. (Bradford) *35*, 1980, 83–88.

PICHI-SERMOLLI, R. E. G., Una carta geobotanica dell'Africa Orientale (Eritrea, Ethiopia, Somalia), Webbia *13*, 1957, 15–132, 1 map.

POPOV, G. / ZELLER, W., Ecological survey report on the 1962 survey in the Arabian peninsula, United Nations Special Fund Desert Locust Project, Progress Report No. UNSF/DL/ES/6, ii + 99 pp.

RATHJENS, C. / WISSMANN, H. VON, Landeskundliche Ergebnisse, Rathjens- und von Wissmannsche Südarabienreise Bd. 3, Hamburgische Universität, Abhandlungen auf dem Gebiet der Auslandskunde *40*, 1934, Hamburg.

RATHJENS, C. / RATHJENS, C. / SAMLESKI, E. / KERNER, G., Beiträge zur Klimakunde Südwest-Arabiens: Das Klima von Sana, das Klima von Jemen, Deutscher Wetterdienst Seewetteramt, Einzelveröffentlichungen, Nr. *11*, 1956, Hamburg.

SCHWARTZ, O., Flora des tropischen Arabien, Mitt. Inst. Allg. Bot. Hamburg *10*, 1939, 1–393.

THESIGER, W., A journey through the Tihama, the 'Asir, and the Hijaz mountains, Geographical Journ. *110*, 1947, 188–200, 1 map.

TROLL, C., Wüstensteppen und Nebeloasen im südnubischen Küstengebirge, Zeitschr. Ges. Erdkunde Berlin, 1935, 241–281.

TROLL, C. / SCHOTTENLOHER, R., Ergebnisse wissenschaftlicher Reisen in Äthiopien. I. Von Eritrea nach Gondar, Addis Abeba und Harar, Peterm. Geogr. Mitt. *85*, 1939, 217–238.

VESEY-FITZGERALD, D. F., Vegetation of the Red Sea coast south of Jedda, Saudi Arabia, Journ. Ecol. *43*, 1955, 477–489, pls. 6–12.

VESEY-FITZGERALD, D. F., The vegetation of the Red Sea coast north of Jedda, Saudi Arabia, Journ. Ecol. *45*, 1957 a, 547–562, pls. 16–19.

VESEY-FITZGERALD, D. F., The vegetation of central and eastern Arabia, Journ. Ecol. *45*, 1957 b, 779–798, 2 fig. pp., pl. 32.

WALTER, H., Einführung in die Phytologie, Bd. III, Grundlagen der Pflanzenverbreitung, 1. Teil, Standortslehre, 1960, Stuttgart.

WALTER, H., Die Vegetation der Erde, Bd. 1, Die tropischen und subtropischen Zonen, 1973, Stuttgart.

WHITE, F., The Afromontane region, In: WERGER, M. J. A. (Ed.), Biogeography and ecology of southern Africa *1*, 1978, 463–513, The Hague.

WICKENS, G. E., The flora of Jebel Marra (Sudan Republic) and its geographical affinities, Kew Bull. Add. Series V, 1976, Royal Botanic Gardens, Kew.

WISSMANN, H. VON, Die *Juniperus*-Gebirgswälder in Arabien. Ihre Stellung zwischen dem borealen und tropisch-afrikanischen Florenreich, In: Geoecology of the high-mountain regions of Eurasia, Proc. Symp. Int. Geogr. Union, Commission on High-Altitude Geoecology, Erdwiss. Forschung *4*, 1972, 157–176, Wiesbaden.

ZOHARY, M., Geobotanical foundations of the Middle East, 2 vol., 1973, Stuttgart & Amsterdam.

Peter König, Institut für Systematische Botanik und Pflanzengeographie, Freie Universität Berlin, Altensteinstr. 6, D-1000 Berlin 33.

Photo 1 *Acacia-Commiphora* xeromorphic woodland, 75 km E al-Qunfuḏa, 500 m (Profile 2).

Photo 2 Sclerophyllous scrub, 25 km NNW al-ʿUlayya, 1900 m (Profile 5).

Photo 3 *Juniperus excelsa* woodland, Ǧabal Saudā', 15 km WNW Abhā, 2850 m (Profile 3).

Photo 4 Semi-desert, 15 km N Abhā, 2100 m (Profile 4).

Ulrich Deil

Die Wadivegetation der nördlichen Tihāma und Gebirgstihāma der Arabischen Republik Jemen

Inhaltsverzeichnis

1. Einleitung

In der spontanen Vegetation des jemenitischen Tieflands spielen azonale Vegetationseinheiten eine wichtige Rolle. Neben der *Avicennia*-Mangrove an der Rotmeerküste und Halophytengesellschaften (mit *Suaeda fruticosa*, *Salsola spinescens*, *Odyssea mucronata*, *Aeluropus lagopoides* u. a.) in unmittelbarer Küstennähe ist dies v. a. die wadibegleitende Vegetation. Sie säumt die großen Wadis, welche aus dem jemenitischen Randgebirge kommend auf die Küstenebene, die Tihāma, austreten. Der großräumige Wassertransport vom niederschlagsreichen Gebirgsanstieg in das semiaride Gebirgsvorland führt dort zu Bodenwasserverhältnissen, die unabhängig von den Niederschlägen der Tihāma selbst sind. Auf diesen, durch Grund- und Oberflächenwasser begünstigten Standorten wachsen Pflanzenbestände, die sich floristisch und physiognomisch von den lichten Trockengehölzen und den Horstgrasfluren der nicht von den Wadis beeinflußten Gebiete abheben. Diese wadibegleitende Vegetation soll hier mit Hilfe von schematisierten Wadiquerprofilen dargestellt werden. Es wird den Zusammenhängen zwischen Vegetation, Relief, Sediment, Gefälle und Abflußdynamik nachgegangen. Da die aktuelle Vegetation Gegenstand der Untersuchung ist, werden auch anthropogene Einflüsse, z. B. Landnutzungsmuster, mit einbezogen.

2. Das Arbeitsgebiet

Fünf große Talsysteme, die ihren Ausgang z. T. schon in den Becken des Hochlandes nehmen, entwässern die Westabdachung des jemenitischen Randgebirges und treten auf die vorgelagerte Küstenebene aus. Von Norden nach Süden sind dies: Wādī Maur, Wādī Surdud, Wādī Sihām, Wādī Rimaʿ und Wādī Zabīd. Die Untersuchung beschränkt sich auf die nördliche Tihāma und Gebirgstihāma der Arabischen Republik Jemen (zwischen 15° und 16° n. Br.). Das Arbeitsgebiet liegt am unteren Mittellauf und Unterlauf des nördlichsten dieser Wadisysteme, am Wādī Maur. Dieses nimmt seinen Ausgang im Norden des Landes im Raum Ṣaʿdah und vereinigt sich kurz vor dem Austritt aus dem Randgebirge mit dem Wādī Lāʿa, das aus dem Randgebirge westlich des Beckens von ʿAmrān kommt. Die Lage des Untersuchungsgebietes zeigt Abb. 1. Es umfaßt einen Ausschnitt aus dem nördlichen Teil der jemenitischen Tihāma und der östlich anschließenden Gebirgstihāma: Diesem Naturraum gehören die nur mäßig geneigten Unterhänge, Talweitungen und Becken des Randgebirges bis in etwa 1000 m Höhe an.

Relief und Substrate:

Einen Eindruck vom Relief und dem Gewässernetz des untersuchten Raumes gibt Abb. 2. Die im Bereich des Wādī Maur etwa 60 km breite Küstenebene steigt von der Rotmeerküste bei al-Luḥayya über az-Zuhra bis zu den ersten Vorbergen östlich der Hauptstraße al-Ḥudaida – Ǧīzān allmählich auf etwa 250 m Meereshöhe an. Es handelt sich um eine flachwellige Aufschüttungsebene, die in ihrem westlichen und zentralen Abschnitt zum Teil von äolischen Sedimenten überlagert ist. Vor allem nördlich von az-Zuhra sind langgezogene, ost-west-streichende, feinsandige Dünen. Die fluviatilen Sedimente sind im gebirgsnahen Abschnitt vorwiegend Kies und Grobsand, im mittleren Abschnitt Feinsand und Schluff. Im küstennahen Bereich wechseln Salztonebenen mit mobilen Dünenfeldern.
Aus der Gebirgstihāma ist im gewählten Ausschnitt vor allem das Einbruchbecken von aṭ-Ṭūr mit seinen Randhöhen erfaßt. Im Beckenboden und an den Unterhängen der Beckenumrahmung liegt das Grundgebirge

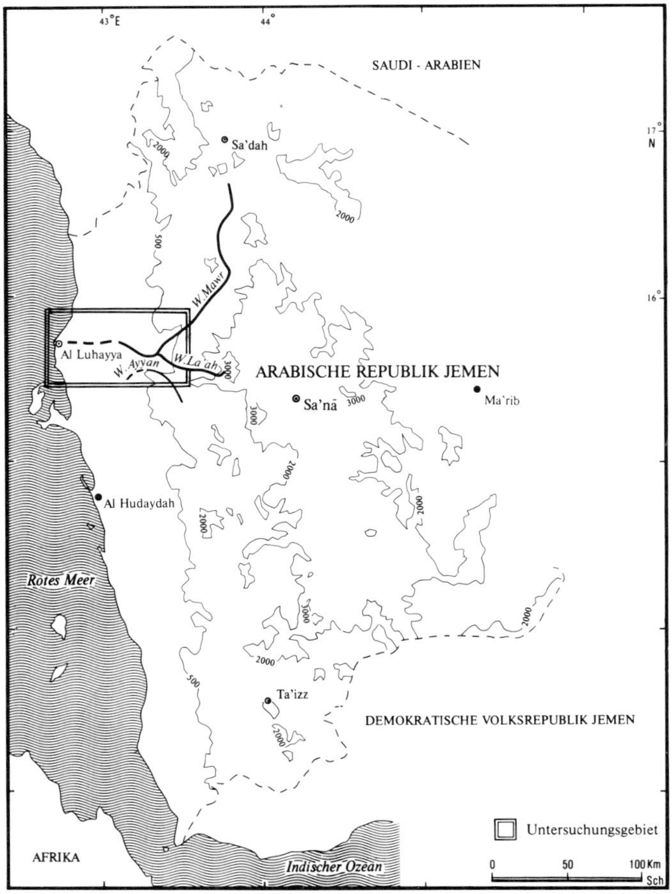

Abb. 1: Lage des Untersuchungsgebietes im Nordjemen

frei (grobkristalline Granite und Gneise und deren Zersatzprodukte), im Norden sind präkambrische Schiefer beteiligt. Im Süden wird das Becken durch eine steil abfallende Bruchstufe begrenzt, die im Ǧabal Nimra und Ǧabal Ǧūba bis auf 1500 m aufsteigt. Ab 800 m ü. NN steht dabei das Deckgebirge an (im Liegenden Sandsteine der Kochlan-Serie, im Hangenden Kalke der Amran-Serie). Im Beckenboden dominieren Fußflächen, die von WNW nach OSO von 250 m auf 500 bis 600 m ansteigen. Sie sind im Ostteil des Beckens stark zerschnitten und in eine Granitkuppenlandschaft aufgelöst.

Klima:

Das Gebiet liegt im Bereich der wechseltrockenen Randtropen. Es ist großklimatisch durch den Wechsel zwischen monsunalen Einflüssen im Sommerhalbjahr und passatischen Einflüssen im Winterhalbjahr gekenn-

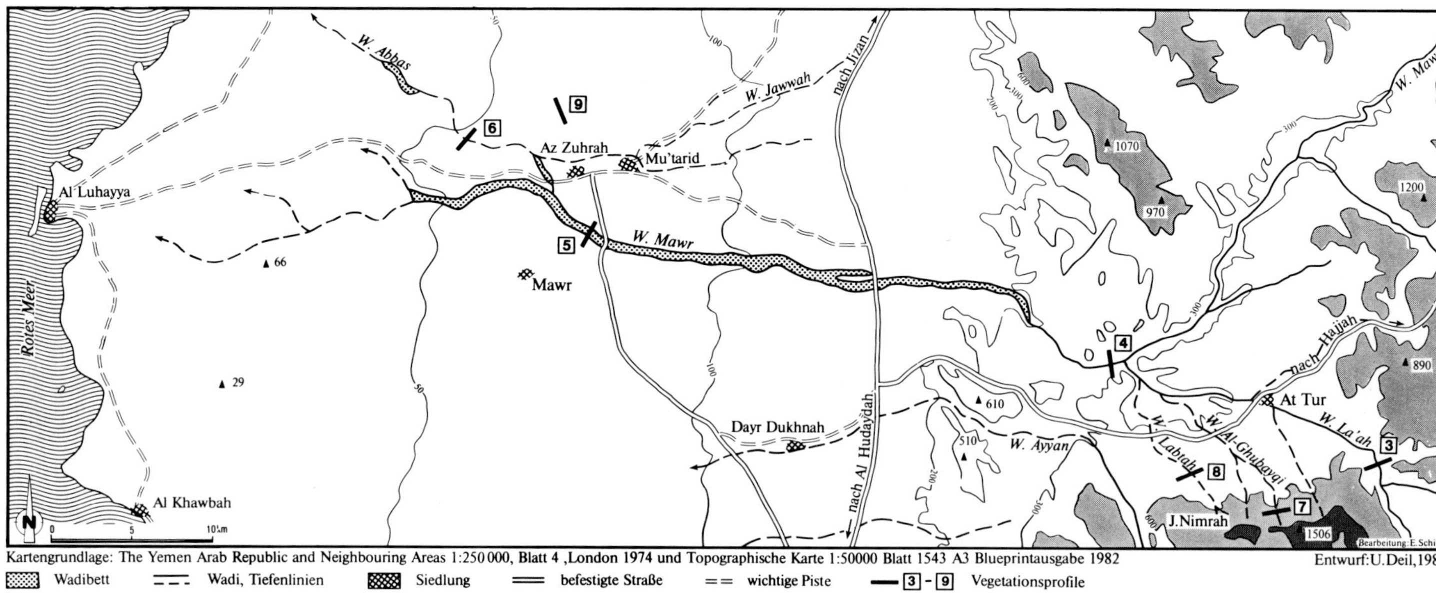

Abb. 2: Das Arbeitsgebiet: Relief, Ortsbezeichnung, Lage der Vegetationsprofile

zeichnet. Zu den unregelmäßig auftretenden und lokal begrenzten Niederschlägen, die meist als nachmittägliche Starkregen fallen, tragen auch lokale Windsysteme und eigenbürtige Zyklonen aus dem Rotmeergraben bei (Flohn 1965).

Zum effektiven Klima liegen keine verläßlichen Daten vor; es können nur Anhaltspunkte gegeben werden. Der Niederschlagsgang in diesem semiariden Gebiet hat zwei Jahresmaxima, ein kleineres im Frühjahr und ein ausgeprägtes im Spätsommer. Die durchschnittlichen Jahressummen der Niederschläge steigen von unter 100 mm an der Küste über etwa 100 mm bei az-Zuhra auf über 200 mm am Gebirgsfuß an. In der Gebirgstihāma liegen sie zwischen 200 und 400 mm. Das Jahresmittel der Lufttemperatur liegt in der Tihāma über 30° C, in der Gebirgstihāma bei etwa 25° C. Beide Gebiete sind ganzjährig frostfrei. Die mittlere Jahresschwankung liegt bei 7–8° C, die Tagesschwankung der Temperatur ist in Rotmeernähe geringer als die Jahresschwankung, in der Gebirgstihāma liegt sie bei 13–14° C. Es herrscht fast stets eine hohe Luftfeuchtigkeit. Für die Pflanzen ist der starke nächtliche Taufall von Bedeutung. In der Tihāma treten tagsüber häufig starke Staubstürme auf.

Gewässernetz und Abflußverhältnisse:

Wegen der insgesamt geringen Niederschläge und wegen ihrer hohen zeitlichen und räumlichen Variabilität schwanken die Abflußverhältnisse auch in den großen Wadis außerordentlich. Ein Kartenbild zeigt daher stets nur einen Augenblickszustand. In Abb. 2 wurde deshalb darauf verzichtet, die Dauer der Wasserführung mit anzuzeigen. Andererseits erreicht ein relativ großer Anteil der Niederschläge die Vorfluter, da wegen der hohen Intensität der Niederschläge und der geringen Infiltrationsrate der Oberflächenabfluß beträchtlich ist.

Perennierend sind nur die beiden Fremdlingsflüsse Wādī Maur und Wādī Lā'a in der Gebirgstihāma. Im gebirgsnahen Teil der Tihāma führen sie noch periodisch Wasser. Eine solche Wasserführung haben ferner das Wādī Ayyān in der Gebirgstihāma und einige autochthone Wadis der Gebirgstihāma in ihren Mittel- und Unterläufen. In allen übrigen Tiefenlinien fließt episodisch Wasser, so im Wādī Maur-Unterlauf im küstennahen Teil der Tihāma, in Spülmuldentälchen der Tihāma und in den Bachschluchten der Oberläufe autochthoner Wadis der Gebirgstihāma. Die Wasserführung der Fremdlingsflüsse wird ganz entscheidend von Niederschlagsereignissen bestimmt, die außerhalb des Untersuchungsgebietes liegen. Ähnlich wie bei der Verteilung der Niederschläge über das Jahr so zeigt sich auch bei den Hochwasserfluten (Sayl) neben einer schwächeren Frühjahrsflut im April/Mai eine kräftigere Sommerflut im August/September. Nur bei extremen Hochwasserfluten fließen Restwassermengen bis zum Roten Meer.

Die Grundwasservorräte der Gebirgstihāma sind sehr begrenzt auf geringmächtige Akkumulationskörper der autochthonen Wadis sowie auf mächtigere Aufschüttungen des Wādī Ayyān im westlichen Beckenteil. Die Grundwasservorräte der Tihāma dagegen sind beträchtlich und liegen in geringer Tiefe (5–25 m). Sie werden durch neu entstandene Brunnen und Pumpenbewässerung zunehmend genutzt und übernutzt. Der Grundwasserstrom in der Tihāma ist zur Küste hin gerichtet. Eine ständige Grundwasserneubildung findet nur aus der Gebirgstihāma her statt, daneben gelegentlich auch nach einem Sayl in der Tihāma, nicht dagegen aus den Niederschlägen der Tihāma selbst (KRAFT et al. 1971).

Talformen und Gefälle:

In der Gebirgstihāma nehmen die meisten Wadis ihren Ursprung in der steilen Randstufe als Bachschluchten oder Kerbtäler, die ins Anstehende eingeschnitten sind. Sie treten mit einem deutlichen Gefälleknick auf den Beckenboden aus und bilden dort nur wenig in die Fußfläche eingetiefte Kerbsohlentäler. Diese dann stark Schotter- und Grus-erfüllten allochthonen Wadis erreichen die Fremdlingsflüsse als Vorfluter meist nicht, sondern enden mit schwachen Schwemmkegeln im Beckeninneren.
Die allochthonen Wadis sind in der Gebirgstihāma mehrere Meter in die Fußflächen eingetieft. In dem kastenförmigen Wadibett verlagert sich der Wasserlauf ständig um Schotterinseln und bildet randlich schmale Akkumulationsterrassen. Das Gefälle des Wādī Lāʿa beträgt beim Eintritt ins Becken von aṭ-Ṭūr über 13‰, bei Einmündung in das Wādī Maur unter 7‰. In der Tihāma hat das Wādī Maur nur noch ein sehr geringes Gefälle (2–3‰). Beim Austritt aus dem Gebirge ist das Wādī Maur noch drei bis vier Meter in die Alluvionen eingetieft, bei az-Zuhra sind es noch 1–2 m. Weiter westlich ist das Wadibett dann morphologisch nicht mehr zu erkennen. Oberflächlichen linienhaften Abfluß nach Niederschlagsereignissen gibt es in der Tihāma nur episodisch in flachen Spülmuldentälchen zwischen den Längsdünen.

Landnutzung:

Das aktuelle Vegetationsmuster wird neben abiotischen Standortfaktoren entscheidend von direkten und indirekten (z. B. über den Wasserhaushalt) Eingriffen des Menschen bestimmt. Die Bevölkerung in diesem Raum lebt auch heute noch fast aussschließlich von der Landwirtschaft. Im Überblick ist dieser Bereich für den gesamten Jemen bei KOPP (1981) dargestellt. Eingehender werden das Becken von aṭ-Ṭūr bei KOPP (1984) und das Wādī Maur-Gebiet um az-Zuhra bei ESCHER (1976) behandelt.
In der Gebirgstihāma stellen Weide und Holzsammelwirtschaft den Haupterwerbszweig dar. Daneben wird „ewiger Hirseanbau“ mit *Sorghum*

bicolor betrieben. Ein gesicherter Ackerbau ist nur bei Zuschußbewässerung möglich. Dabei treten zwei Bewässerungsformen auf:

- Die Sawagi-Bewässerung. Dem Feld wird der Oberflächenabfluß einer benachbarten Regensammelfläche zugeleitet, die meist einen degradierten Trockenwald trägt. Diese Form überwiegt im Beckeninneren und an den Unterhängen auf Spül- und Fußflächen.
- Bei der Sayl-Bewässerung (= Flutbewässerung) wird die Hochwasserwelle aus dem Wadi durch Ableitdämme benachbarten Feldkomplexen zugeleitet. Dieses Bewässerungsprinzip wird vor allem an den kleineren autochthonen Wadis betrieben. Die großen Wadis sind zu tief eingeschnitten. Hier beschränkt sich die Flutbewässerung auf schmale, flußbegleitende Terrassen.

In der Tihāma ist der Ackerbau der Haupterwerbszweig. Dabei ist die Saylbewässerung aus dem Wādī Maur und dem Wādī Ayyān die Grundlage der Bewässerungslandwirtschaft. Als „Wadi“ wird das gesamte mit Flußwasser bewässerbare Areal bezeichnet (Escher 1976), nicht nur das eigentliche Wadibett. In der Trockenzeit werden in Gemeinschaftsarbeit im Wadibett Ablenkdämme (= aqm) aus Erde und Schotter errichtet. Sie tragen eigene Namen und unterstehen einem Aufseher. Das Wasser wird auf Wallfelder geleitet, die bis zu 1 m Höhe überstaut werden. Nach dem Einsickern wird gepflügt und eingesät. Der Wasservorrat im Boden reicht dann für eine Ernte. Im Wādī Maur gilt, wie in den meisten Wadis, das Recht des Oberliegers. Dennoch bietet nicht die gebirgsnahe Zone die besten Bedingungen. Wegen des starken Gefälles werden die Dämme oft weggerissen, außerdem führt das Wadi dort noch sehr grobe Schotter. Begünstigt ist die mittlere Zone etwa von der Ǧīzān-Straße bis az-Zuhra. Dieser Bereich wird von schwachen und kräftigen Fluten erreicht, so daß 2–3 Ernten eingebracht werden können. Weiter westlich reicht der Sayl nur gelegentlich. Dort lagert sich fruchtbares Feinsediment ab, das zu hohen Erträgen führt. Die Nutzfläche schwankt aber außerordentlich stark von Jahr zu Jahr und es sind großflächig Brachen verbreitet. Im Bewässerungsfeldbau werden vor allem Durah *(= Sorghum bicolor)* angebaut, daneben im Winter auch Baumwolle. Das nicht bewässerbare Land („Khabt“), vor allem die konsolidierten Sanddünen und die gebirgsnahen Spülflächen, werden im Regenfeldbau bewirtschaftet. Bei ca. 150 mm Niederschlag ist die Ertragsunsicherheit extrem groß, so daß nicht in jedem Jahr mit einer Ernte gerechnet werden kann (Kopp 1981). Hier wird in vermehrtem Umfang Duhn *(= Pennisetum glaucum)* auf Langstreifenfluren im Pflug- und Hackbau angebaut.

3. Die Wadi-Vegetation

Anhand einiger Wadiquerprofile werden die Zonierung der Wadisaumgesellschaften und ihre Kontaktvegetation dargestellt. Die Auswahl der Profile ist repräsentativ, d.h., es werden alle wichtigen Vegetationstypen und Gerinnebettformen erfaßt. Die Darstellung beschränkt sich auf dominante und kennzeichnende Arten. Eine eingehende pflanzensoziologische Bearbeitung findet sich bei DEIL/MÜLLER-HOHENSTEIN (1985a). Die Nomenklatur der Arten folgt DEIL/MÜLLER-HOHENSTEIN (1985b).
Die Lage der Profile im Relief und Gewässernetz des Arbeitsgebietes zeigt Abb. 2. Zunächst wird mit vier Profilen (Abb. 3–6) die Vegetation an den allochthonen Wadis von der Gebirgstihāma in die Tihāma hinaus verfolgt. In einem zweiten Teil wird mit drei weiteren Profilen (Abb. 7–9) auf die Vegetation an den autochthonen Wadis eingegangen. Für alle Profile gilt, daß sie stets flußabwärts gesehen sind. Der gewählte Ausschnitt und der Maßstab wechseln stark. Die schematisierten Bestandsaufrisse sollen einen Eindruck von der Schichtung und den dominierenden Lebensformen der Gesellschaften geben. Bei immergrünen Arten ist die Krone dunkel dargestellt, bei breitlaubig-mesomorphen, trockenkahlen Arten mittelgrau, bei mikrophyllen trockenkahlen Arten hellgrau.

3.1 Die Vegetation an großen, allochthonen Wadis

3.1.1 Allochthone Wadis in der Gebirgstihāma

Die Situation an den großen Wadis der Gebirgstihāma zeigen die Abb. 3 und 4. Das erste Profil liegt südöstlich von aṭ-Ṭūr beim Eintritt des Wādī Lāʿa in das Becken in etwa 450 m Höhe. Der Fluß hat hier ein Gefälle von fast 14‰. Das zweite Profil liegt am Nordwestrand des Beckens nach der Einmündung des Wādī Lāʿa in das Wādī Maur in etwas über 200 m Höhe. Das Gefälle hat sich auf ca. 7‰ verringert. In beiden Fällen ist das Gerinnebett 5–10 m tief kastenförmig eingetieft, z.T. ins Anstehende, das an den Prallhängen zutage tritt, meist jedoch in einen Akkumulationskörper aus wechsellagernden Grobschottern, Kiesen und Sanden. Sowohl das Gerinnebett als auch das Tal weiten sich nach unten zu (siehe hierzu auch Bild 1 und 2). Höherliegende, flußbegleitende Terrassen gehen allmählich in weite Spülflächen über. In beiden Abschnitten führen die Flüsse nahezu ständig Wasser und Hochwasserfluten sind häufig. Die Wassermenge ändert sich beim Durchfluß durch die Gebirgstihāma nur wenig, da eine Ableitung der Hochwasserwellen auf das Ackerland wegen der tief eingeschnittenen Gerinne nicht möglich ist. Beide Flußabschnitte sind durch Materialdurchtransport und -umlagerung sowie durch eine häufige Verlagerung der Abflußrinnen gekennzeichnet. Abnehmendes Gefälle und

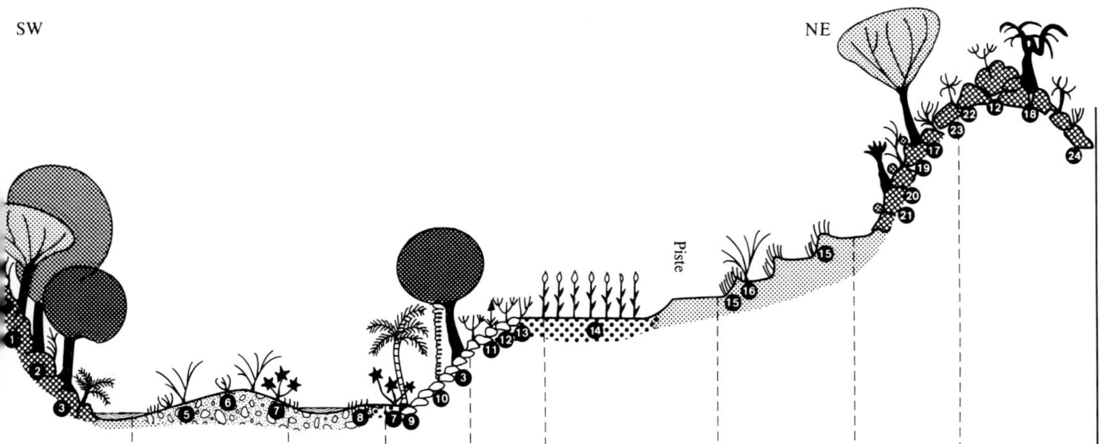

lbmesophytischer ckenwald	Trockengehölz-Initialen	Aktuelles Gerinnebett und Uferkrautflur	Meso-phytischer Galeriewald	Sukkulentenflur an Wüstungsplatz	Ackerland	Mähwiesen	Trockengehölz	Sukkulentenflur
Prallhang auf tehendem Gestein anit, Schiefer) mit *Acacia asak,* *ortilis* s.str., *nellifera* s.str. *Tamarindus indica* *Berchemia discolor,* *chilia emetica* s.str., *nbretum molle* *Abrus bottae,* *nocarpum yemenense,* *vtenus senegalensis,* *rissa edulis*	auf übersandeter Grobschotterinsel mit ❺ *Acacia ehrenbergiana,* selten auch *A.tortilis* ❻ *Cassia senna* ❼ *Jatropha curcas*	mit ❽ *Bacopa monnieri, Fimbristylis cymosa, Phyla nodiflora, Eclipta alba* u.a.	am Fuß des Gleithanges ❾ *Phoenix reclinata* ❸ *Berchemia discolor, Trichilia emetica* s.str. *Combretum molle* ❼ *Jatropha curcas,* selten *Annona squamosa* ❿ Lianenschleier mit *Cissus quadrangularis, C. rotundifolia* u.a.	auf Grobschotterterrasse ⓫ mit *Aloe vera officinalis* ⓬ *Euphorbia inarticulata* ⓭ *Caralluma penicillata, C. subulata*	auf grobsandig-grusiger Terrasse, sawagi-bewässert, Feldfrüchte: ⓮ *Sorghum bicolor,* als Nebenfrucht *Sesamum indicum* und *Vigna aconitifolia* Unkrautgesellschaft mit *Cyperus rubicundus, C. rotundus, Heliotropium* div.spec. *Pulicaria jaubertii,* u.a. Kurzfristige Brachen mit *Solanum dubium, Aerva javanica, Jatropha lobata glauca* u.a.	auf feinsandigen Dämmen mit ⓯ *Heteropogon contortus, Bothriochloa insculpta, Cenchrus pennisetiformis, C. setigerus* und anderen ausdauernden Gräsern bei Brachfallen der Terrassenfläche: Vorwaldstadien mit ⓰ *Acacia ehrenbergiana*	am Hang im Granitblockschutt mit ⓱ *Acacia asak,* selten *A. tortilis* ⓲ *Commiphora abyssinica simplicifolia,* ⓳ *Abrus bottae Anisotes trisulcus, Premna resinosa, Maytenus senegalensis* ⓴ *Adenium obesum, Adenia venenata* ㉑ *Indigofera spinosa, I. oblongifolia, Seddera latifolia, S. virgata*	auf Kuppen in grusiger Spaltenfüllung zwischen anstehendem Granit, mit ㉒ *Aloe sabaea* ⓬ *Euphorbia inarticulata* ㉓ *Kleinia odora* ㉔ *Sansevieria ehrenbergii, Eulophia petersii, Caralluma penicillata,* u.a.

wurf: U. Deil 1982

b. 3: Vegetationsprofil durch das Wādī Lāʿa vor Eintritt in das Becken von aṭ-Ṭūr/Gebirgstihāma
(schematisiert und stark überhöht)

geringere Strömungsgeschwindigkeit haben zur Folge, daß im Abschnitt des Profils 3 Grobschotter und Kies liegenbleiben, während weiter flußabwärts Kies- und Sandinseln und sandige Niederterrassen entstehen. Dies wirkt sich unmittelbar auf die Wadivegetation aus.

Auf höhergelegenen, konsolidierten Schotterinseln bildet die Pionierbaumart *Acacia ehrenbergiana* zusammen mit *Cassia senna* die Initialgesellschaft zu den Trockengehölzen hin. Meist gesellt sich noch *Acacia tortilis* hinzu. Die Entwicklung bleibt jedoch wegen der Umlagerung auf diesem Stadium stehen. Im Wādī Maur entwickelt sich auf dem feineren Sediment im Flußbett eine Tamarisken-Initialgesellschaft mit *Tamarix nilotica.* Den Ufersaum besiedelt eine Hochgrasflur mit der herdenbilden-

den *Desmostachya bipinnata* und vereinzelten Büschen von *Tamarix nilotica* und *Jatropha curcas. Desmostachya* stabilisiert durch ihre Horste und unterirdischen Ausläufer die sandige Niederterrasse. Sie vermag der wiederkehrenden Überflutung und Übersandung durch Höhenverlagerung der Sproßachsen zu folgen. Auch die Weiterentwicklung dieser Dauergesellschaft wird durch die Abflußdynamik verhindert. Nur in ruhigen Seitenbuchten kann sich ein geschlossener Tamarisken-Auwald ausbilden. An den *Desmostachya*-Saum schließt sich uferferner auch hier die *Acacia ehrenbergiana-Cassia senna*-Gesellschaft an. Den wasserzügigen Fuß der Hochterrasse säumt *Salvadora persica.*
Der *Desmostachya bipinnata-Jatropha curcas*-Saum ist meist nur an den Gleithängen besser ausgebildet (Bild 2). Ein solcher ist in Abb. 4 erfaßt. Welchen Einfluß ein anderes Substrat im Einzugsgebiet des Flusses auf die Ufervegetation hat, zeigt ein Vergleich von Wādī Ayyān und Wādī Maur im Becken von aṭ-Ṭūr: Beide haben beim Austritt aus der Gebirgstihāma gleiches Gefälle. Auch am Wādī Ayyān findet man *Tamarix nilotica-* und *Acacia ehrenbergiana-Cassia senna*-Gehölze, nicht jedoch die *Desmostachya-Jatropha curcas*-Gesellschaft. Im Einzugsgebiet dieses Wadis stehen überwiegend Kalke des Deckgebirges an. Es führt daher nur Schotter und schüttet keine Sandterrassen auf.
In beiden Flußabschnitten entwickeln sich nach Hochwasserwellen kurzzeitige Krautfluren. Sie werden von der Scrophulariacee *Bacopa monnieri,* anderen niederliegenden Dikotylen und ausläuferbildenden Süß- und Sauergräsern aufgebaut. Die *Bacopa*-Gesellschaft liegt nur einige Dezimeter über der Wadisohle und wird daher auch bei schwachem Sayl überflutet.
Im Bereich des Profils 3 wird das Wadi durch einen mesophytischen Galeriewald charakterisiert, der von breitlaubig-mesomorphen, trockenkahlen Arten *(Berchemia discolor, Trichilia emetica, Combretum molle, Tamarindus indica)* und der immergrünen *Phoenix reclinata* gebildet wird. Im Unterwuchs haben sich *Jatropha curcas* und *Annona squamosa* eingebürgert. Am flachgründigen Prallhang treten Elemente der lichten Trockengehölze hinzu (*Acacia* div. spec., die Fabaceen *Abrus bottae* und *Ormocarpum yemenense*).
Artenreiche xerophytische Trockengehölze bilden die Schlußgesellschaften an den Hängen. Hochstete Kennarten dieser heute meist mehr oder weniger stark degradierten Bestände sind *Acacia tortilis, Commiphora opobalsamum, C. abyssinica, C. myrrha, Adenium obesum, Boscia augustifolia, Anisotes trisulcus, Abrus bottae, Ormocarpum yemenense, Seddera-* und *Indigofera*-Arten. In tieferen Lagen (250–500 m) treten in der *Acacia mellifera-Cadaba glandulosa*-Gesellschaft neben den namengebenden Arten *Cadaba rotundifolia* und *Delonix elata* auf, in höheren Lagen (von 450 m an aufwärts) *Acacia asak, Premna resinosa, Maytenus senegalensis* u. a.

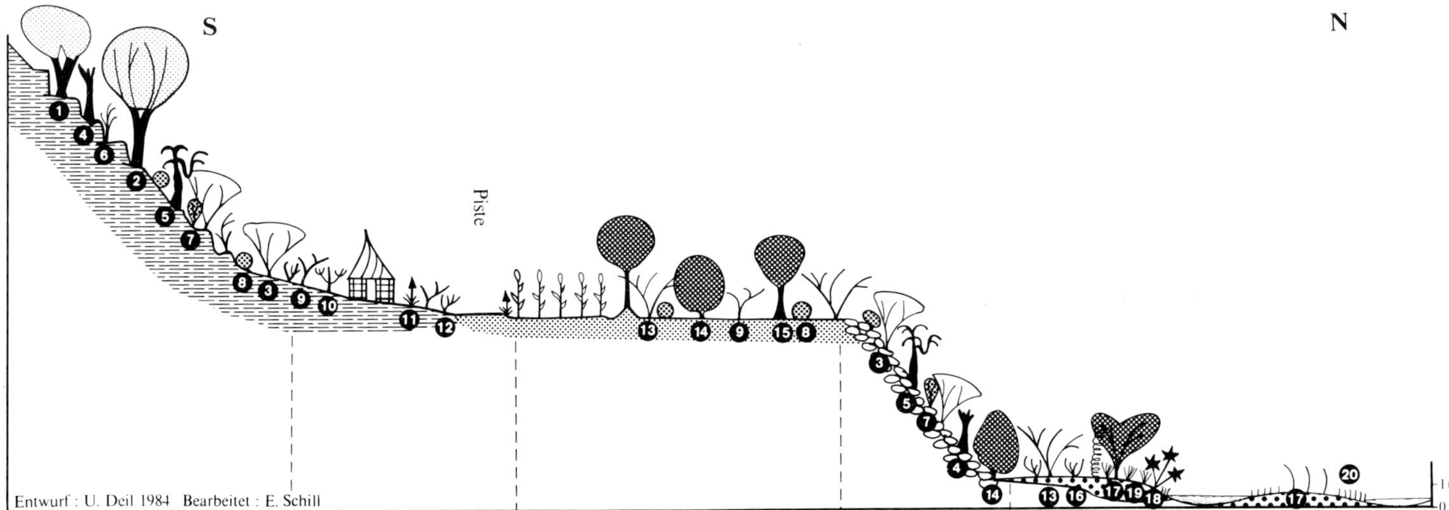

Xerophytischer Trockenwald am Hang auf Gneis- und Schieferschutt mit	**Sukkulentenflur** am Hangfuß an Dorfrändern und auf Wüstungsplätzen mit	**Halbimmergrünes Trockengehölz** auf sandig-schluffiger Spülfläche mit	**Xerophytisches Trockengehölz** am Hochterrassen- oder Fußflächenhang, stark degradiert mit	**Halbimmergrünes Trockengehölz und Hochgrasflur** auf sandiger Niederterrasse; uferfern mit:	**Tamariskeninitialen und Uferkrautflur** auf Sandinseln mit
1 *Acacia mellifera* 2 *Delonix elata* 3 *Acacia tortilis* s.str. 4 *Adenium obesum* 5 *Commiphora opobalsamum*, *C.abyssinica simplicifolia* 6 *Abrus bottae* 7 *Anisotes trisulcus* 8 *Cadaba glandulosa*, *C.rotundifolia*	9 *Acacia oerfota* 10 *Euphorbia inarticulata* 11 *Aloe vera officinalis* 12 *Caralluma penicillata*, *C.subulata* u.a. *Stapelieae*	13 *Acacia ehrenbergiana* 14 *Salvadora persica* 15 *Dobera glabra* 8 *Cadaba glandulosa*, *C.rotundifolia* 9 *Acacia oerfota* meist stark aufgelichtet oder gerodet für Hirseanbau *(Sorghum bicolor)*	3 *Acacia tortilis* s.str., *A.mellifera* s.str. (Stockausschläge) 4 *Adenium obesum* 5 *Commiphora* div.spec. 7 *Anisotes trisulcus*; am Hangfuß ein Saum mit 14 *Salvadora persica*	13 *Acacia ehrenbergiana* 16 *Cassia senna*; ufernahe mit: 17 *Tamarix nilotica* 18 *Jatropha curcas* 19 *Desmostachya bipinnata*	17 *Tamarix nilotica* - Jungpflanzen 20 *Bacopa monnieri*, *Fimbristylis cymosa*, *Phyla nodiflora*, *Eclipta alba* u.a.

Abb. 4: Vegetationsprofil durch das Wādī Maur beim Austritt aus der Gebirgstihāma (schematisiert und stark überhöht)

An besonders flachgründigen Standorten (Kuppen, Terrassenkanten) und an solchen, die einem extremen Weidedruck ausgesetzt sind (Dorfrandlagen, Wüstungsplätze) kommen verschiedene sukkulentenreiche Gesellschaften vor (mit *Aloe* div. spec., *Euphorbia inarticulata, Caralluma* div. spec., *Acacia oerfota* u. a.).
Die ackerbauliche Nutzung ist im Durchbruchstal des Wādī Lāʿa durch die Umrahmung des Beckens von aṭ-Ṭūr auf kleine Bereiche begrenzt (Bild 1). Die Unterhänge und die Talmulden von Seitentälchen sind zu beetartigen Terrassen umgestaltet. Dort werden die Terrassenkronen und -hänge von einem *Heteropogon contortus*-Grasland bestanden, das als Mähwiese genutzt wird. Die Terrassenfläche selbst ist heute oft aufgelassen. *Acacia ehrenbergiana* bildet hier ein Vorwaldstadium zum *Acacia asak*-Trockenwald hin. Die hochgelegenen, flußbegleitenden Terrassen erhalten Zuschußwasser von den Hängen (Sawagi-Bewässerung) und werden mit *Sorghum bicolor*-Hirse bestellt. Das reichliche Wasserangebot im Wadi kann meist nicht genutzt werden. Nur ganz kleinflächig können dem Flußbett Terrassenleisten abgerungen werden, die von Hochwässern immer wieder bedroht und zerstört werden. Hier können die Felder ganzjährig bewässert und sehr intensiv bewirtschaftet werden (Fruchtgärten mit Papaya, Bananen, Annonen; Gemüse- und Maisanbau; Dauergrünland).
In den Beckenlagen und beim Austritt der Wadis aus der Gebirgstihāma gehen die Hochterrassen in ausgedehnte Fußflächen und Schwemmebenen über. Die mit einem Steinpflaster bedeckten Fußflächen werden meist als Regensammelflächen und Kleinviehweiden genutzt und tragen sehr stark degradierte *Acacia mellifera*-Trockengehölze. Die mit mächtigeren Sedimenten bedeckten Spülflächen sind für die ackerbauliche Nutzung besser geeignet. Die dort ursprünglich vorhandenen halbimmergrünen Trockengehölze mit *Acacia ehrenbergiana, A. oerfota, Dobera glabra* und *Salvadora persica* verschwinden heute zunehmend. Steht ein mächtiger, gut grundwasserleitender Akkumulationskörper zur Verfügung, so findet man hier auch im Bereich der Gebirgstihāma moderne Brunnen und Pumpenbewässerung.

3.1.2 Allochthone Wadis in der Tihāma

Mit dem Austritt aus der Gebirgstihāma verändern sich Gefälle, Abflußverhalten und Gerinnebettform der allochthonen Wadis. Zwei Profile durch das Wādī Maur im Raum az-Zuhra sollen beispielhaft die Situation im Unterlauf zeigen. Profil 5 liegt kurz unterhalb der Piste az-Zuhra – al-Qanāwiṣ, Profil 6 nordwestlich az-Zuhra an einem auslaufenden Nebenarm des Wādī Maur (Wādī Abbās), der dort an Dünenfelder grenzt (damit Anschluß an die Geländesituation im Profil 9). Das Gefälle hat sich auf

2–3 ‰ verringert. Die Hochwasserfluten erreichen az-Zuhra regelmäßig, das Wādī Abbās nur gelegentlich im Sommer. Sie führen nur noch feine Sedimente mit sich. Die Strömungsgeschwindigkeit hat sich so weit verringert, daß ausdauernde, hochwüchsige Pflanzengesellschaften das Wadibett selbst besiedeln können (siehe auch Bild 3 u. 4). Bei az-Zuhra ist das Wadibett ca. 2–4 m gegenüber dem Bewässerungsland eingetieft, am Wādī Abbās sind es nur noch 1–2 m. Weiter westlich tritt das Wadibett morphologisch nicht mehr hervor. Tamariskenbestände zeigen aber den Grundwasserstrom an.

Im Abschnitt zwischen Ǧīzān-Hauptstraße und az-Zuhra findet eine so umfangreiche Wasserableitung statt, daß der Sayl weitgehend aufgezehrt wird. Die Ablenkdämme halten hier stand. Sie leiten das Wasser in Flutrinnen (siehe Abb. 5), welche das Wadibett zunächst mit geringerem Gefälle begleiten und dann auf die Wallfelder abzweigen.

Das Pflanzenkleid ist in diesem Abschnitt durch das Vorkommen von *Saccharum spontaneum* gekennzeichnet. Dieses bis 2 m hoch wachsende Gras bildet im Wadibett dichte Horste, welche den Fluten standhalten. Dazwischen sind Tamariskenbüsche (*Tamarix nilotica* und *T. aphylla*). Gelegentlich sind auch *Jatropha curcas* und *Parkinsonia aculeata* eingebürgert. Nur an dichteren Stellen vermag sich zwischen den *Saccharum*-Horsten auch *Desmostachya bipinnata* anzusiedeln und bildet dann eine zweite Krautschicht. Nach dem Hochwasser entstehen kurzzeitige Krautfluren mit niederliegenden Kräutern *(Eclipta alba, Glinus lctoides)* und ausläuferbildenden Süß- und Sauergräsern *(Brachiaria reptans, Dactyloctenium scindicum, Cyperus rotundus, Fimbristylis cymosa)*. Weiter flußabwärts (Profil 6) fehlt *Saccharum spontaneum*. Hier besiedeln Tamarisken und *Salvadora persica* das Wadibett. Etwas höher gelegene Terrassen tragen die *Acacia ehrenbergiana-Cassia senna*-Gesellschaft. Dieser Abschnitt wird nur noch gelegentlich von sehr starken Hochwasserfluten erreicht. Im Wadibett lagert sich tonig-schluffiges Material ab. Während der Trockenzeit entstehen sehr starke Trockenrisse und es findet eine oberflächliche Salzanreicherung statt. Nach dem Sayl entsteht kurzzeitig ein Rasen mit dem sukkulenten Annuellen *Zygophyllum simplex*. Westlich der 50 m-Isohypse wächst dann bereits *Suaeda fruticosa* zusammen mit *Tamarix* in den dann kaum noch erkennbaren Tiefenlinien.

Etwas weiter als im Wadibett können sich die Gesellschaften am Hochufer entwickeln. Dort sind z.T. noch recht dichte Bestände mit *Salvadora persica*, *Tamarix* div. spec. und *Jatropha curcas* erhalten. Den Unterwuchs bilden *Desmostachya bipinnata*-Herden. Diese Art besiedelt auch die Flutrinne, die von Gehölzen freigehalten wird. Sie greift von dort aus in die angrenzenden Felder über. Durch das Pflügen, durch Abflammen oder durch Schnitt zur Gewinnung von Fasermaterial wird sie immer wieder zurückgedrängt. In dem großflächigen Bewässerungsland ist die Spontan-

Aktuelles Gerinnebett

grob – feinsandig
Tamarisken – Auwaldstreifen und *Saccharum* – Initialen auf Sandinseln mit
❶ *Saccharum spontaneum* var. *aegyptiacum*
❷ *Tamarix nilotica* und *T. aphylla*
im Schutz dichterer Bestände seltener
❸ *Desmostachya bipinnata*
nach sayl: kurzzeitige Krautflur mit
❹ *Cyperus rotundus, Fimbristylis cymosa, Brachiaria reptans, Dactyloctenium scindicum, Eragrostis minor, Corchorus* div.spec. *Eclipta alba, Flaveria trinervia, Glinus lotoides* u.a.

Hochufer

feinsandig – schluffig
Gehölzstreifen mit
❺ *Salvadora persica,*
❻ *Jatropha curcas*
❷ *Tamarix nilotica* und *T. aphylla*
dazwischen Hochgrasflur mit ❸ *Desmostachya bipinnata*
Lianenschleier mit
Leptadenia arborea, Cissus quadrangularis, Momordica balsamina, Cocculus pendulus

Flutrinne
(Bewässerungskanal)

feinsandig – schluffig
Hochgrasflur mit
❸ *Desmostachya bipinnata*
(oft geschnitten oder abgeflammt)
vereinzelt Sträucher
❼ *Indigofera oblongifolia*
❽ *Abutilon pannosum, Senra incana*

Wallfelder

schluffig – tonig, oberflächlich verdichtet, sayl – bewässert
Kulturpflanzen : ❾ überwiegend *Sorghum bicolor,* seltener *Pennisetum glaucum, Sesamum indicum,* als Nebenfrucht oft *Vigna aconitifolia*
häufige Unkräuter : *Pulicaria jaubertii, Heliotropium longiflorum, Corchorus* div.spec. *Cleome* div.spec, *Dactyloctenium aegyptium, Paspalidium desertorum, Cynodon dactylon, Cyperus rotundus* u.a.
kurzfristige Brachen mit
❿ *Zygophyllum simplex* – Rasen (kurzlebiger Therophyt)
⓫ *Schouwia purpurea, Tephrosia purpurea* (längerlebige Therophyten)
⓬ ausdauernde Sträucher wie *Aerva javanica, Cassia italica, C. holosericea, Jatropha lobata glauca, Solanum dubium, Heliotropium* div.spec.
längerfristige Brachen mit ❼ *Indigofera oblongifolia*
❽ *Abutilon pannosum, Senra incana*, an den Dämmen meist *Tamarix nilotica,* seltener *Ziziphus spina-christi* oder *Ficus sycomorus;* in Dorfnähe auch *Moringa oleifera* (gepflanzt)

Entwurf : U. Deil
Bearbeitung: E. Schill

Abb. 5: Vegetationsprofil durch den Hauptast des Wādī Maur südöstlich az-Zuhra/Tihāma (schematisiert und überhöht)

vegetation weitgehend verschwunden. Nur vereinzelt findet man als Schattenbäume auf den Dämmen *Tamarix nilotica*, *Ziziphus spina-christi* oder *Ficus sycomorus*. Hier dominiert heute großflächig der Hirseanbau, meist mit *Sorghum bicolor*, seltener mit *Pennisetum glaucum*. Die Felder enthalten eine artenreiche Unkrautflora. Da das Wasser nie für alle Parzellen reicht, findet man hier auch stets Brachestadien. Auf dem schluffig-tonigen Substrat im Bewässerungsland beginnt die Sukzession mit *Zygophyllum simplex*-Rasen. In einigen Wochen entwickeln sich dann höherwüchsige und längerlebige Therophyten wie *Schouwia purpurea* und *Tephrosia purpurea*. Bei längerfristigem Brachfallen stellen sich ausdauernde Sträucher ein (*Cassia* div. spec., *Heliotropium* div. spec., *Solanum dubium*, *Indigofera oblongifolia* und die beiden Malvaceen *Abutilon pannosum* und *Senra incana*).
Auch im „Khabt“, dem nicht vom Zuschußwasser des Wadis beeinflußten Land, sind die dort ursprünglich vorhandenen lichten Trockengehölze mit *Acacia tortilis*, *A. ehrenbergiana*, *A. oerfota*, *Maerua crassifolia* und *Leptadenia pyrotechnica* heute weitgehend verschwunden.
Auf den festliegenden Dünenzügen nördlich az-Zuhra (siehe Abb. 6 und Bild 6) sind jetzt Kleinstrauch- und Horstgrasgesellschaften großflächig verbreitet. Bei den Horstgräsern dominiert *Panicum turgidum*, daneben treten auch *Lasiurus hirsutus* und *Pennisetum divisum* auf. Bei den Sträuchern sind *Salsola spinescens*, *Blyttia spiralis* und *Jatropha glandulosa* häufig. Auch das Khabt-Land wird ackerbaulich genutzt. Nach ausreichenden Regenfällen wird Duhn *(= Pennisetum glaucum)* eingesät. Kurzfristige Brachen auf sandigem Substrat sind durch *Dipterygium glaucum* gekennzeichnet. In allen Pflanzengesellschaften auf den Dünen entwickelt sich nach Regenfällen ein Acheb, eine Annuellenflur mit *Indigofera* div. spec., *Blepharis ciliaris*, *Leptothrium senegalense* u.a.

3.2 Die Vegetation an kleinen, autochthonen Wadis und in Spülmulden

Die geringen Niederschläge und das Sayl-Bewässerungsprinzip führen dazu, daß im Untersuchungsgebiet keine perennierenden Wasserläufe ihren Ausgang nehmen. Einige wenige, schwach schüttende Quellen werden zur Bewässerung kleiner Hausgärten aufgezehrt. Die Wasserversorgung von Mensch und Tier wird aus oberflächennahem Grundwasser sichergestellt. Es gibt aber eine Vielzahl von episodisch bis periodisch wasserführenden Tiefenlinien, da unter den heutigen Klimabedingungen die linienhafte Zerschneidung überwiegt. Die Talformen hängen vor allem vom Gefälle ab, weniger von Untergrund und Größe des Einzugsgebietes. Mit zwei Querprofilen (Abb. 7 und 8) sollen die autochthonen Wadis der Gebirgstihāma und ihre Vegetation dargestellt werden, mit einem Profil (Abb. 9) die in der Tihāma auftretenden Spülmulden.

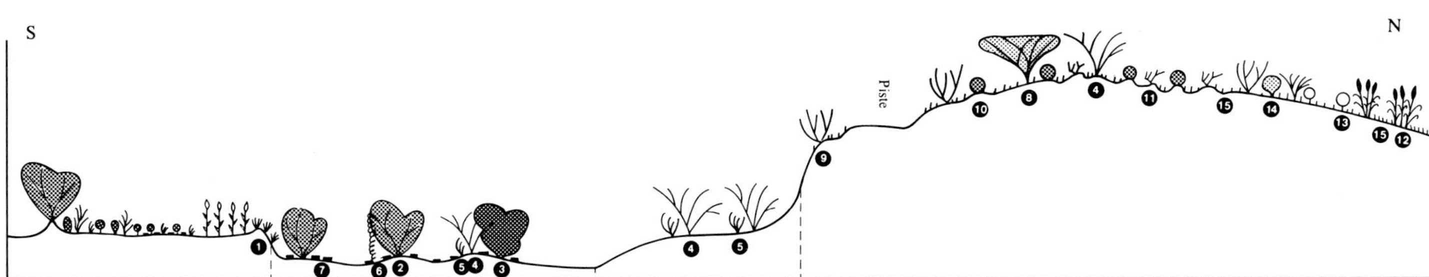

Wallfelder

saylbewässert, tonig – schluffig, oberflächlich verdichtet
Vegetation wie bei Wallfeldern in Abb. 5 (Signaturen siehe dort), aber Brachestadien hier häufiger, da die Felder seltener überflutet und bestellt werden
auf den Wällen : *Tamarix* div.spec., selten *Capparis decidua*
auf wadinahen Wällen und Dämmen:
(1) *Desmostachya bipinnata*

Aktuelles Gerinnebett

tonig – schluffig
Immergrüne Gehölze mit
(2) *Tamarix aphylla* und *T. nilotica*
(3) *Salvadora persica*
selten auch
(4) *Acacia ehrenbergiana*
(5) *Cassia senna*
(6) Lianenschleier mit *Cissus quadrangularis, Momordica balsamina*
nach Flutwelle (= sayl) :
(7) *Zygophyllum simplex* – Rasen

Terrasse

tonig – schluffig, z.T. übersandet
Trockengehölz mit
(4) *Acacia ehrenbergiana*
(5) *Cassia senna*

Dünenzug mit Primärdünen - Kleinrelief

mittel – feinsandig, am Unterhang zunehmend schluffig ; Düne insgesamt ruhend, v.a. im Bereich des Dünenkammes aber kleinräumige äolische Umlagerung
Trockengehölzreste (z.B. an Friedhöfen) mit
(8) *Acacia tortilis* s.str. (4) *Acacia ehrenbergiana*, selten *A. oerfota*
(9) *Leptadenia pyrotechnica*
Zwergstrauch / Horstgrasgesellschaft mit
(10) *Jatropha glandulosa, Salsola spinescens, Blyttia spiralis, Indigofera spinosa*
(11) *Panicum turgidum*
Ackerland mit Regenfeldbau ; als Feldfrucht (12) *Pennisetum glaucum*
kurzfristige Brachen mit (13) *Dipterygium glaucum, Tephrosia purpurea, Heliotropium pterocarpum, H. strigosum* u.a.
längerfristige Brachen mit (14) *Lasiurus hirsutus*, selten auch *Acacia ehrenbergiana* – Jungpflanzen
nach Regenfällen kurzfristiger Acheb mit (15) *Indigofera hochstetteri, I. argentea, I. sessiliflora, Crotalaria microphylla, Euphorbia scordifolia, Boerhavia coccinea, Blepharis ciliaris, Odontanthera radicans, Leptothrium senegalense* u.a.

Entwurf U. Deil

Abb. 6: Vegetationsprofil durch den nördlichen Seitenarm des Wādī Maur (Wādī ʿAbbās) nordwestlich az-Zuhra/Tihāma
(schematisiert und stark überhöht)

3.2.1 Autochthone Wadis in der Gebirgstihāma

Im Becken von aṭ-Ṭūr nehmen die meisten der autochthonen Wadis ihren Ausgang in den Steilhängen der Beckenumrahmung. In den höheren Lagen (450–900 m) des südlichen Beckenrandes ist ein eigener Wadibett- und Wadivegetationstyp ausgebildet (Abb. 7). In die steilen, mit 30 bis 40° nach Nord geneigten Hänge des Ǧabal Nimra sind Kerbtälchen eingetieft, in denen ein mesophytischer *Combretum molle*-Schluchtwald gedeiht. Das Bachbett ist nur wenige Meter breit und ins Anstehende (Grundgebirgsgranit) eingeschnitten. Dieses liegt offen zutage oder ist von einer nur geringmächtigen Grusauflage oder gerundeten Felsblöcken bedeckt. Nach Regenschauern tritt sehr kurzzeitig (stundenweise) ein heftiger oberfläch-

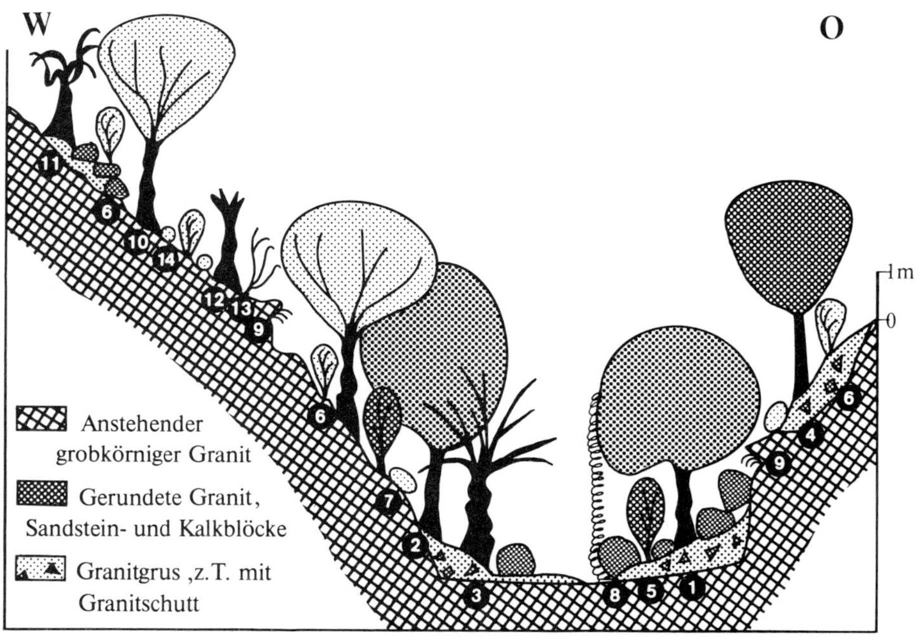

Schluchtwald im Bacheinschnitt mit ❶ *Combretum molle, Berchemia discolor* ❷ *Tamarindus indica* ❸ *Ficus* div.spec. ❹ *Boscia angustifolia,* selten *Dobera glabra, Mimusops laurifolia* ❺ *Ochna inermis, Tarenna graveolens arabica, Ehretia obtusifolia, Dodonaea viscosa, Maytenus* div.spec., *Carissa edulis* ❻ *Premna resinosa, Ecbolium linnaeanum, Grewia* div.spec., *Anisotes trisulcus* ❼ *Barleria* div.spec. ❽ *Cissus rotundifolia, Cocculus hirsutus , Cassytha filiformis* ❾ Poikilohydre Farne und Moosfarne
Trockenwald am Hang mit ❿ *Acacia asak* ⓫ *Commiphora abyssinica, C.myrrha* ⓬ *Adenium obesum* ⓭ *Abrus bottae, Ormocarpum yemenense* ⓮ *Seddera* div.spec.
❻ und ❾ siehe oben

Abb. 7: Vegetationsprofil durch eine Bachschlucht im Oberlauf des Wādī l-Ġubaiqī/Gebirgstihāma

licher Abfluß ein. Es überwiegt der Materialab- und durchtransport. Das Bachbett selbst ist weitgehend vegetationsfrei, der Wald stockt auf den steilen Talflanken.

In diesem Talabschnitt gibt es keinen Grundwasserkörper, von dem die bachbegleitende Vegetation profitieren könnte. Sie erhält jedoch das oberflächliche Zuschußwasser und ist, lokalklimatisch bedingt, nur geringen Transpirationsverlusten ausgesetzt. Insbesondere während der Trockenzeit heben sich die dunkelgrünen Schluchtwälder von den graubraunen Trockengehölzen am Hang ab. Die Baumschicht wird von regengrünen, breitlaubig-mesomorphen *(Combretum molle, Berchemia discolor, Tamarindus indica)* und immergrünen, breitlaubig-xeromorphen Arten *(Boscia angustifolia, Mimusops laurifolia, Ficus salicifolia, Dobera glabra)* gebildet. Einige weitere *Ficus*-Arten *(F. glumosa, F. sycomorus, F. lutea, F. populifolia)* werfen ihr Laub in langen Trockenzeiten ab.

Im Unterwuchs treten hartlaubige, immergrüne Sträucher hervor. Sie haben ihren Schwerpunkt in den regen- und nebelreichen höheren Lagen des Randgebirges, reichen in diesen Bachschluchten jedoch weit in die Gebirgstihāma herab (*Ochna inermis, Tarenna graveolens* ssp. *arabica, Carissa edulis, Ehretia obtusifolia, Maytenus* div. spec.). Dies gilt auch für die halbkugeligen Dornsträucher *Barleria hochstetteri* und *B. bispinosa.* Daneben kommen trockenkahle Sträucher vor, die auch in den *Acacia asak-Commiphora*-Trockengehölzen am offenen Hang zu finden sind (*Abrus bottae, Oromocarpum yemenense, Premna resinosa* und verschiedene *Grewia*-Arten). Im Schatten überhängender Felsen und in sickerfrischen Spalten gedeiht eine Synusie aus poikilohydren Gefäßkryptogamen (*Adiantum incisum, A. capillus-veneris, Actiniopteris semiflabellata* und *Selaginella imbricata*). An länger überrieselten Stellen können sich Kalktuffkissen mit kleinflächigen Naturwiesen entwickeln. Die dominanten Arten sind *Heteropogon contortus, Arthraxon prionodes* und *Fimbristylis cymosa.*

Der *Combretum molle*-Schluchtwald hat viele gemeinsame Arten mit dem mesophytischen Galeriewald an größeren Wadis, wie er in Profil 3 beschrieben wurde (z. B. *Combretum molle, Berchemia discolor* und *Tamarindus indica*). Es fehlen aber *Trichilia emetica* und *Phoenix reclinata.* Dafür ist der Unterwuchs reicher an immergrünen Arten. Die Bestände sind heute nur noch in Relikten vorhanden und durch Wasserableitung, Holzeinschlag und Schneiteln stark im Rückgang begriffen.

In Nordlagen unterhalb 450 m, in Südexpositionen schon unter 600 m wird der *Combretum molle*-Schluchtwald von einem artenreichen, xeromorphen Trockengehölz abgelöst. Als Kennart tritt hier *Acacia abyssinica* in der Varietät *macroloba* auf, daneben ist die gesamte Artengarnitur der *Acacia-Commiphora*-Trockengehölze vertreten. Da die Bestandesstruktur und die Ökologie der Gesellschaft schon ausführlich dargestellt wurden (Deil/Müller-Hohenstein 1985 a, insb. Abb. 4), hier nur einige Anmer-

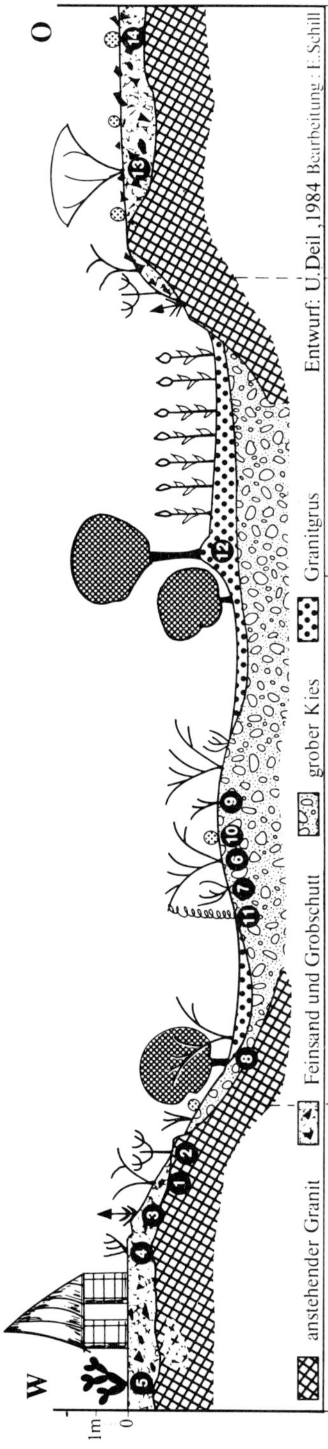

Sukkulentenflur
an Fußflächenkante um Dörfer und an Wüstungsplätzen mit
1 *Acacia oerfota*
2 *Euphorbia inarticulata*
3 *Aloe vera officinalis*
4 *Caralluma* div.spec.
5 *Opuntia dillenii* (gepfl.)

Halbimmergrünes Auengehölz im Wadibett mit
6 *Acacia ehrenbergiana*, slt. auch *A.tortilis* s.str.
7 *Cassia senna*
8 *Salvadora persica*
9 *Acalypha fruticosa*, *Aerva javanica*, *Jatropha lobata glauca*, *Ecbolium linnaeanum*
10 *Indigofera spinosa*, *I. oblongifolia*
11 *Cissus rotundifolia* u.a. Lianen

Saylbewässerte Hirsefelder
an den Dämmen als Schneitelbäume
12 *Dobera glabra*, slt. *Ziziphus spina-christi*
gelegentlich auch pumpenbewässert, dann Fruchtbaum- und Gemüsegärten mit *Acalypha indica*- *Heliotropium longiflorum* - Fazies

Xerophytisches Trockengehölz
stark degradiert, Fußfläche als Regensammelfläche, mit
13 *Acacia mellifera* s.str.
A. tortilis s.str.
14 *Indigofera* div.spec.,
Seddera div.spec.

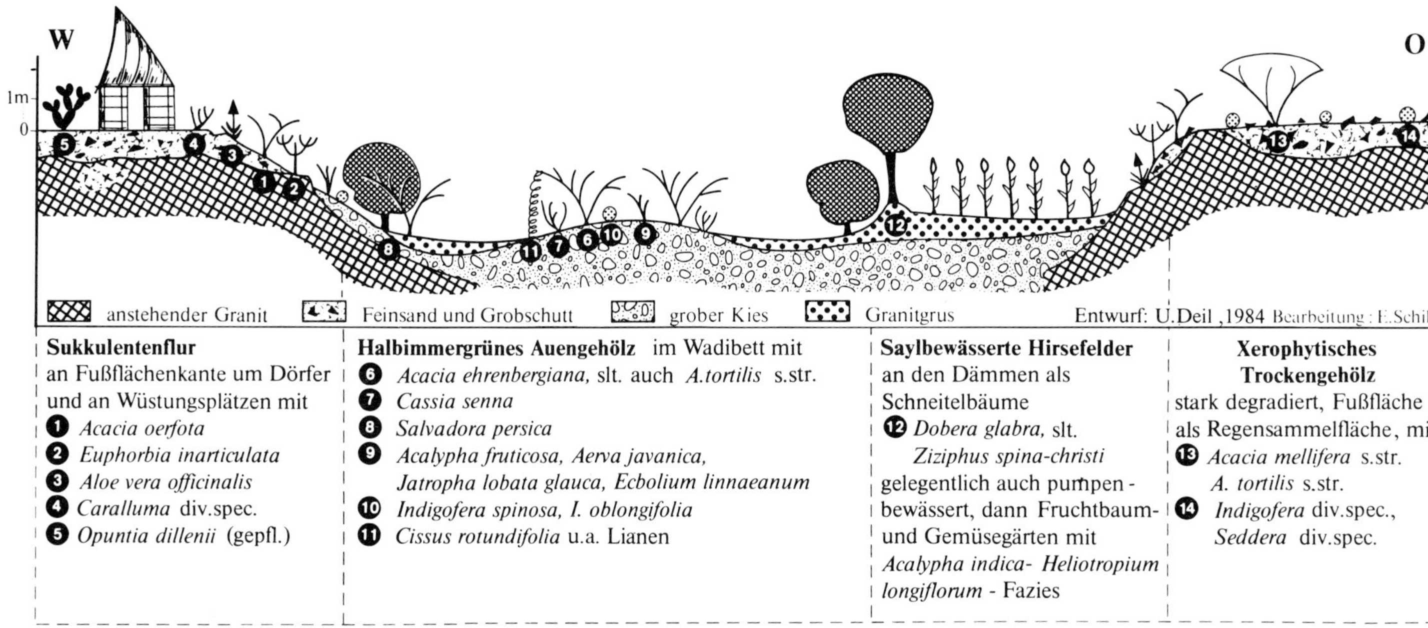

Abb. 8: Vegetationsprofil durch den Mittellauf des Wādī Labṭah/Becken von aṭ-Ṭūr/Gebirgstihāma

kungen: Die *Acacia abyssinica*-Gesellschaft findet man in Kerbsohlentälchen, welche in die Unterhänge der Beckenumrahmung oder auf die Fußflächen in den anstehenden Granit eingeschnitten sind. Die Abflußverhältnisse sind wie im *Combretum molle*-Schluchtwald: episodisch starker Durchfluß; kein Grundwasserstrom; schmale, vegetationsfreie Bachsohle; die Bäume und Sträucher siedeln in Klüften an den steilen Talflanken.
Mit einem deutlichen Gefälleknick treten die Wadis von der Randstufe auf die Fußflächen im Inneren des Beckens von aṭ-Ṭūr aus (Bild 5). Schlagartig ändern sich damit Talform, Abflußbedingungen und Wadivegetation. Das Talbett wird sehr breit, ist schottererfüllt und nur wenige Meter in die Fußfläche eingetieft (Abb. 8). Die starke Sedimentation hat ihre Ursache nicht nur im geringeren Gefälle, sondern in der Ableitung des Sayls. Wo immer möglich, werden dem anastomosierenden Flußbett Terrassenkomplexe abgerungen. Die ursprüngliche Vegetation im Wadibett ist die *Acacia ehrenbergiana-Cassia senna*-Gesellschaft (siehe auch Abb. 3). Am Rand des Wadis, im Mündungslauf auch im Bachbett selbst, ist *Salvadora persica* beteiligt. Die Sträucher im Unterwuchs haben z.T. ruderalen Charakter *(Acalypha fruticosa)* oder finden sich auch auf Brachen (*Aerva javanica*, *Jatropha lobata* ssp. *glauca*). Die Gesellschaft unterliegt einem starken Weidedruck. *Acacia ehrenbergiana* und *Salvadora persica* werden gerne von Kamelen gefressen. *Cassia senna* dagegen ist ungenießbar. Die Kontaktvegetation sind zum einen die Unkraut- und Brachegesellschaften der Hirsefelder im Wadibett, zum anderen extrem degradierte *Acacia mellifera*-Trockengehölze auf den Fußflächen und Sukkulentenfluren an den Kanten der Fußflächen. Die letztgenannte Reliefsituation stellt bevorzugte Siedlungsplätze dar. Die Wadis werden oft auch bei starkem Sayl im Beckeninneren aufgezehrt und enden in flachen Schwemmfächern mit einer *Salvadora persica*-Fazies. Gelegentlich erreichen sie jedoch die allochthonen Wadis.

3.2.2 Spülmulden in der Tihāma

Im Gebiet des Bewässerungsfeldbaus („Wadi“) entstehen in der Tihāma auch bei heftigen Niederschlägen keine Gerinne, da der Regen dort auf eingeebnetes und von Wällen umgebenes Land fällt. Im Regenfeldbaugebiet („Khabt“) gibt es jedoch Tiefenlinien mit episodischem Abfluß. Es sind flache Spülmuldentälchen (Bild 6). Die Infiltrationsrate auf den sandigen Dünen ist zwar relativ hoch, dennoch kommt es gelegentlich zu Oberflächenabfluß. Daneben ist in den Dünentälern auch ein geringer Grundwasserstrom vorhanden, der vereinzelt für einige Jahre mit Pumpen ausgebeutet wird. Die Brunnen fallen aber bald wieder trocken und die Felder werden aufgegeben. Im Normalfall wird in diesem Gebiet von Halbnomaden nach Regenfällen Duhn angebaut und eine extensive Weidewirtschaft betrieben.

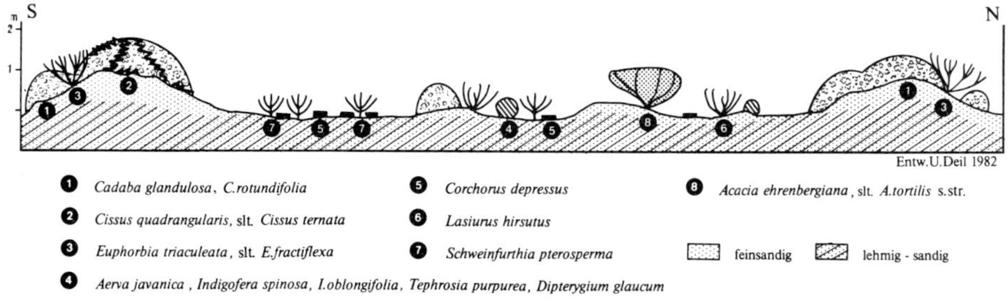

Abb. 9: Vegetationsprofil durch ein Dünental mit *Cadaba*-Nebket nördlich az-Zuhra/Tihāma

Für die Dünentäler ist die *Cadaba-Schweinfurthia pterosperma*-Gesellschaft charakteristisch. Die beiden Capparidaceen *Cadaba glandulosa* und *C. rotundifolia* bilden Nebka-ähnliche Sandhügel. In ihrem Schutz wachsen Sukkulente (die schlingenden Vitaceen *Cissus quadrangularis* und *C. ternata* und die Kandelabereuphorbien *Euphorbia triaculeata* und *E. fractiflexa*). Nur selten kommen auch Makrophanerophyten auf *(Acacia ehrenbergiana, A. tortilis, Maerua crassifolia).* Zwischen den Nebket läuft nach dem Regen ein Teppich mit Therophytenkeimlingen auf. Kennzeichnend für das verdichtete, sandig-lehmige Substrat sind *Schweinfurthia pterosperma* und *Corchorus depressus.* Bei starker Durchfeuchtung oder wiederholten Regenfällen sind die Therophyten potentiell auch längerlebig und können dann an der Basis verholzte Halbsträucher bilden *(Tephrosia, Schweinfurthia, Dipterygium).* Die *Cadaba*-Tälchen enden nicht in abflußlosen Hohlformen, sondern laufen mit sehr geringem Gefälle zu einem der Wādī Maur-Arme hin aus. Inwiefern die Nebket Anzeichen für eine Desertifikation sind (Mobilisierung fossiler Dünensande nach Zerstörung der schützenden Vegetationsdecke) müssen weitere Untersuchungen zeigen.

3.3 Zusammenfassung und Typisierung der Wadivegetation

Aus der Beschreibung der natürlichen Bedingungen und der agrarwirtschaftlichen Eingriffe im Untersuchungsgebiet wurde deutlich, daß vom Gefälle direkt die Abflußdynamik und die Sedimentationsbedingungen abhängen und damit indirekt die Gerinnebettformen. Diese und die Strömungsgeschwindigkeit wiederum sind entscheidend für die Möglichkeit der Wasserableitung zur Flutbewässerung. Wie im Querprofil, so gibt es auch im Längsprofil der Wadis in den einzelnen Abschnitten eine spezifische Vegetation und damit insgesamt eine charakteristische Abfolge von Pflanzengesellschaften. Solche Abfolgen werden mit Rudich/Danin (1978) als Sequenzen bezeichnet.

Sequenz 1: Im Verlauf eines autochthonen Wadis der Gebirgstihāma von der Randstufe in den Beckenboden bis hin zum Fremdlingsfluß als Vorfluter folgt auf den *Combretum molle*-Schluchtwaldtyp höherer Lagen der *Acacia abyssinica* var. *macroloba*-Typ tieferer Lagen, im Beckenboden schließlich der *Acacia ehrenbergiana-Cassia senna*-Typ.
Die Spülmulden der Tihāma werden in ihrem mittleren und unteren Abschnitt durch den *Cadaba-Schweinfurthia pterosperma*-Typ charakterisiert. Für eine detaillierte Darstellung in einer Sequenz liegt noch zu wenig Aufnahmematerial vor.
Sequenz 2: Auch im Verlauf der allochthonen Wadis von der Randstufe durch die Gebirgstihāma in die Tihāma hinaus wandelt sich die Wadivegetation gesetzmäßig ab (Abb. 10): Auf den *Breonadia salicina*-Typ der Randstufe folgt zunächst ein Abschnitt in der Gebirgstihāma mit stärkerem Gefälle, der den *Berchemia discolor*-Typ als einen weiteren mesophytischen Galeriewald trägt. Im flacheren Teil der Gebirgstihāma und in der gebirgsnahen Tihāma säumt der *Tamarix-Desmostachya bipinnata*-Typ das Wadibett. Im zentralen Teil der Tihāma ist das Wadibett selbst vom *Tamarix-Saccharum spontaneum*-Typ besiedelt, im küstennahen Teil vom *Tamarix-Zygophyllum simplex*-Typ. An der Rotmeerküste gedeihen *Hyphaene thebaica*-Bestände in grundwassernahen Tiefenlinien. Diese Sequenz

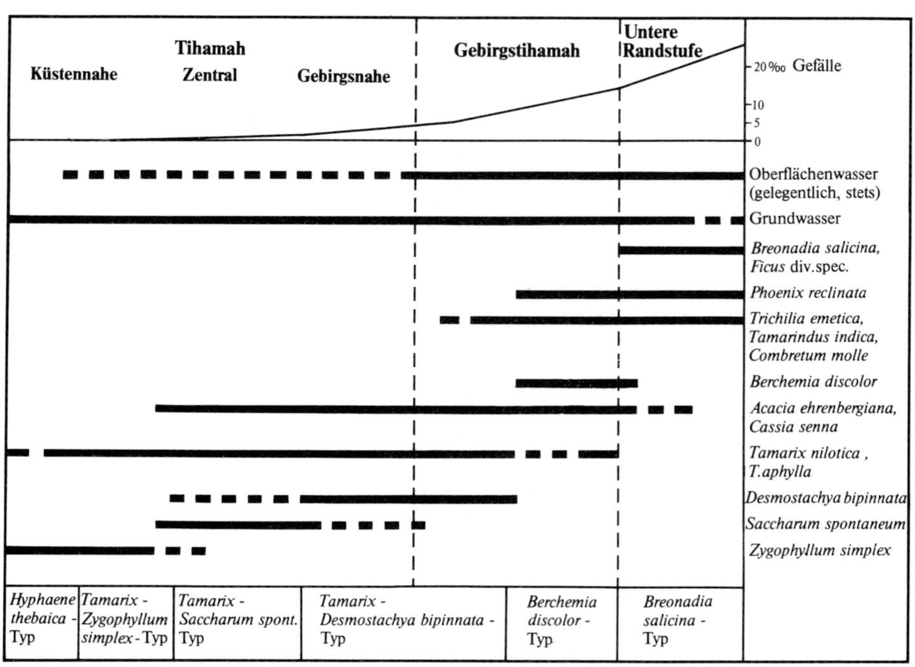

Abb. 10: Vegetationstypen an allochthonen Wadis der Tihāma und Gebirgstihāma

ist noch unvollständig, da die Wadivegetationstypen des Oberlaufes nicht bekannt sind.

Die Vegetationseinheiten am Wadi haben eine Reihe von Gemeinsamkeiten, welche sie von der Vegetation im „Khab“-Land unterscheidet:

a) Sie werden von mesomorpheren Lebensformen als die Schlußvegetation an den Hängen bestimmt.

b) Sie haben zwar nicht in allen Fällen Grundwasser zur Verfügung, erhalten aber stets Zuschußwasser aus dem Oberflächenabfluß.

c) In mehreren Fällen entstehen Dauergesellschaften, deren Weiterentwicklung durch die Abflußdynamik verhindert wird (*Acacia ehrenbergiana-Cassia senna*-Gesellschaft, *Tamarix-Desmostachya bipinnata*- und *Tamarix-Saccharum spontaneum*-Gesellschaft). In anderen Fällen wird hierdurch sogar die Besiedelung des Bachbettes selbst verhindert (*Combretum molle*- und *Acacia abyssinica*-Gesellschaft) (ein Kennzeichen des Quellaufabschnitts auch in Alpenflüssen).

d) Für die Ausbildung von ökologischen Reihen im Querprofil und von Sequenzen im Längsprofil der Wadis scheint weniger die Höhe des Grundwasserstandes und die Dauer der Überflutung eine Rolle zu spielen als vielmehr Abflußdynamik und Korngröße des Sediments. Letztere hängt von der Schleppkraft des Flusses und damit im Querprofil von der Höhenlage über dem Wadi, im Längsprofil vom Gefälle ab. Vergleichbares hat Wendelberger (1973 und 1984) auch für die Auenvegetation Mitteleuropas festgestellt.

e) Die Wasserversorgung und damit die phänologische Entwicklung der Bestände ist (mit Ausnahme der *Combretum molle*- und der *Cadaba-Schweinfurthia*-Gesellschaft) unabhängig von Niederschlagsereignissen im entsprechenden Wadiabschnitt.

f) Eine klimatische Differenzierung des Untersuchungsgebietes zeigt sich aber auch in der Wadivegetation und zwar darin, daß einige Arten, welche im trockenen Bereich an Tiefenlinien gebunden sind, in niederschlagsreicheren Gebieten auch auf die Hänge übergehen (z. B. *Cadaba glandulosa*, *C. rotundifolia*, *Combretum molle*, *Berchemia discolor*, *Acacia abyssinica*, *A. ehrenbergiana* und verschiedene *Ficus*-Arten).

g) Vergleicht man die Arealtypen der Vegetationseinheiten, so nimmt der Anteil von Arten mit eng umgrenzten Vorkommen in der Folge Unkrautgesellschaften – Wadiges. – Trockengehölzges. – Sukkulentenges. zu. Die Wadivegetation als typisch azonale Vegetation enthält viele weit verbreitete Elemente der wechselfeuchten Randtropen Afrikas (z. B. *Tamarindus indica*, *Berchemia discolor*, *Ficus* div. spec., *Acacia ehrenbergiana*, *Cassia senna*) bzw. der altweltlichen Trockengebiete (*Desmostachya bipinnata*, *Salvadora persica*, *Saccharum spontaneum*, *Zygophyllum simplex*) und nur wenige eritreo-arabische Florenelemente (z. B. *Combretum molle*).

Diesen chorologischen Aspekten soll am Beispiel der *Tamarix*-Bestände nachgegangen werden.

4. Ein überregionaler Vergleich: Tamarix nilotica-Bestände

Die Fassung der obengenannten Waditypen in Pflanzengesellschaften, ihre detaillierte floristische Zusammensetzung, synsystematische Stellung und Ökologie wurden für die Gebirgstihāma bereits behandelt (DEIL/MÜLLER-HOHENSTEIN 1985 a). Für den Bereich der Tihāma ist eine entsprechende Studie in Vorbereitung. Deshalb wird hier auf diese Gesichtspunkte nicht näher eingegangen. Am Beispiel der *Tamarix nilotica*-Bestände, welche ein wesentliches Element der Wadigesellschaften in diesem Raum sind, soll jedoch einigen chorologischen Phänomenen nachgegangen werden.

Die Verbreitung von *Tamarix nilotica* zeigt Abb. 11. Hierbei wurden die Vorkommen von *T. nilotica*, *T. mannifera* und *T. arabica* im Sinne von BAUM (1978) zusammengefaßt, da es sich um nur schlecht charakterisierte Kleinarten eines Formenkreises handelt (so verfahren z.B. auch ZOHARY et al. 1983). Das Hauptvorkommen von *Tamarix nilotica* liegt in der eritreo-arabischen Florenprovinz Sensu ZOHARY (1966). Daneben gibt es Ausstrahlungen ins Nildelta, auf den Sinai, nach Jordanien und in trockene Teile Ostafrikas.

Sowohl auf Überblickskarten des Klimas als auch der Vegetation der Erde fällt der Raum beiderseits des Rotmeergrabens und die Somali-Halbinsel aus der üblichen Anordnung in breitenparallelen Zonen heraus. FLOHN (1964) weist daraufhin, daß das Trockengebiet in Nordostafrika durch das Ausbleiben der für diesen Bereich sonst typischen Sommerregen entsteht und erörtert die möglichen Ursachen. Schon RATHJENS/KERNER (1956) weisen ferner auf eine Winterregenenklave am Südausgang des Roten Meeres hin. Floristisch entspricht diesem „Problemklimagebiet“ die eritreo-arabische Florenprovinz. Charakteristische Geoelemente sind z.B. die beiden Euphorbiaceen *Jatropha lobata* ssp. *glauca* und *J. glandulosa*. Auch die Sektion *Diacanthium* der Gattung *Euphorbia* hat hier ein Mannigfaltigkeitszentrum mit vielen endemischen Arten.

Aus diesem Raum sind Vergesellschaftungen von *Tamarix nilotica* beschrieben, wie sie auch im Jemen auftreten, z.B. mit *Desmostachya bipinnata* und *Saccharum spontaneum*. Die *Tamarix*-Gesellschaften stehen mehrfach (so in Nordafrika und im ʿAsir) in Kontakt zur *Acacia ehrenbergiana*-*Cassia senna*-Gesellschaft.

Im Areal von *Tamarix nilotica* vollzieht sich ein bedeutender klimatischer Wandel: Der Übergang von den warmtemperierten Subtropen zur Tropenzone (siehe z.B. die Grenzziehung nach der Klimagliederung von

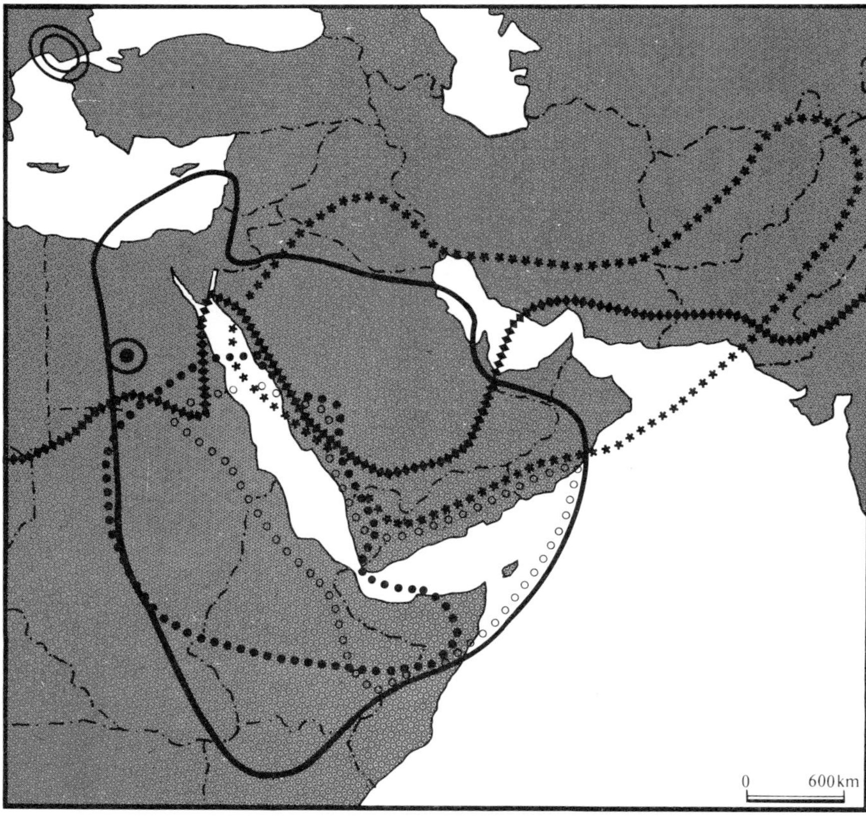

♦♦♦♦♦ Grenze zwischen Tropenzone und warmtemperiert - subtropischer Zone der Klimagliederung nach TROLL/PAFFEN (aus MÜLLER 1982)

— *Tamarix nilotica* - Formenkreis (= *T. nilotica, T. mannifera, T. arabica sensu* BAUM 1978)

••• *Jatropha lobata* ssp. *glauca* ○○○ *J. glandulosa*

*** *Rhazya stricta* ⊙ *R. greissii* ◎ *R. orientalis*

Abb. 11: Ausgewählte Klima- und Arealtypen im eritreo-arabischen Raum

Troll/Paffen, in Müller 1982 Abb. 11). Klimagenetisch ist es der Wechsel vom extratropischen (mediterran- und nordsaharo-arabischen) Winterregengebiet in das randtropische Sommerregengebiet. Im einzelnen wird jedoch die Klimagenese stark differenziert durch die orographischen Verhältnisse im Rotmeergraben (siehe hierzu vor allem Flohn 1965). So ergibt sich eine ganz verschiedene jahreszeitliche Verteilung der Niederschläge für das Hochland, die Randstufe und die Küstengebiete beiderseits des Roten Meeres.

Floristisch drückt sich der klimatische Wandel in diesem Raum im Übergang vom holarktischen zum paläotropischen Florenreich aus. Dieser Florenwandel ist innerhalb der *Tamarix nilotica*-Gesellschaften sehr schön nachzuvollziehen: Im Tiefland am Rotmeergraben bis etwa in Höhe Ǧidda sind *Jatropha*-Arten (*J. lobata* ssp. *glauca*, *J. glandulosa* und die eingebürgerte *J. curcas*) als tropische Genoelemente am Aufbau der *Tamarix*-Gesellschaften beteiligt (Vesey-Fitzgerald 1955 und eigene Beobachtungen). Nördlich von Ǧidda im Küstentiefland (Vesey-Fitzgerald 1957a), im Zentralen Arabien (Vesey-Fitzgerald 1957b und Baierle et al. 1985) (dort z.T. mit *Acacia ehrenbergiana*), im jemenitischen Hochland (Westinga/Thalen 1980 und eigene Beobachtungen) und im Inneren von Hadramaut ist häufig die Apocynacee *Rhazya stricta* am Aufbau der Wadigesellschaften beteiligt.

Die Verbreitung der Gattung *Rhazya* zeigt Abb. 11. Neben der weit verbreiteten *Rhazya stricta* (ein arabisch-sindisches Geoelement) gibt es noch zwei weitere, hochendemische Arten: *R. greissii* in den Kharga-Oasen der Westlichen ägyptischen Wüste (Täckholm 1974) und *R. orientalis* im thrazischen Küstentiefland (ursprünglich auch in Nordwest-Anatolien, nach Tutin et al. 1972). Eine Karte bei Meusel et al. (1978) und Angaben in verschiedenen Florenwerken zeigen, daß *Rhazya stricta* im feuchteren Teil der subtropischen Winterregengebiete durch die Gattung *Nerium* abgelöst wird (so z.B. im mediterran beeinflußten Teil des Vorderen Orients, in höheren Lagen des südlichen Iran, Ostafghanistans und des nördlichen Pakistan). Die beiden Apocynaceen-Gattungen sind ebenfalls tropischer Abstammung. Sie vertreten einander in azonalen Tamariskengesellschaften im Zentrum und am Nordrand des saharo-arabisch-sindischen Trokkengürtels. Daß sie viel weiter als die Gattung *Jatropha* in den extratropischen Raum vorstoßen konnten, hängt mit ihrer Frosthärte zusammen. Auch im Jemen sind *Jatropha glandulosa* und *J. lobata* ssp. *glauca* auf die frostfreie Tieflandstufe beschränkt, im Hochland (z.B. bei Marib) tritt *Rhazya stricta* in den *Tamarix nilotica-Saccharum spontaneum*- und *Tamarix nilotica-Desmostachya bipinnata*-Beständen auf.

Eine synsystematische Gliederung der Wadivegetation der semihumid-semiariden Subtropen und Randtropen Nord- und Ostafrikas sowie des Vorderen und Mittleren Orients steht noch aus. Hierzu können *Tamarix*-Formenkreise, Vertreter der Gattung *Jatropha* und die Gattungen *Rhazya* und *Nerium* herangezogen werden. Eine wichtige Rolle in diesen Gesellschaften spielen ferner strömungs- und überschüttungsfeste Horstgräser *(Desmostachya bipinnata)* und die Gattungen der *Saccharinae (Imperata, Saccharum, Erianthus)* innerhalb der Tribus *Andropogonae.* De Bolos (1968) hat als übergeordnetes Syntaxon für diese Gesellschaften die Klassengruppe Imperato-Tamaricea vorgeschlagen. Er lenkt damit den Blick auf eine auffällige Übereinstimmung der Zonierung an den Wadis des jemenitischen Tieflandes mit folgenden Gebieten:

– Quezel (1965) beschreibt vom Sahara-Südrand den Kontakt einer grundwassernahen *Tamarix brachystylis-Desmostachya bipinnata*- mit einer grundwasserferneren *Acacia nilotica* ssp. *adstringens-Desmostachya bipinnata*-Gesellschaft am Rand abflußloser Seen mit starken Wasserspiegelschwankungen. Sehr anschaulich sind auch die Transekte an den Sebkas in Nordost-Niger von Boudouresque/Schulz (1981). Sie zeigen den Kontakt zwischen Tamariskengehölzen (*T. brachystylis* und *T. aphylla*), Hochgrasfluren (*Imperata cylindrica* und *Desmostachya bipinnata*) und lichten Trockengehölzen (*Acacia ehrenbergiana* und *Cassia senna*). In höheren Lagen der südsaharischen Gebirge (Tibesti und Hoggar) wird die *Tamarix gallica-Desmostachya*-Gesellschaft von einer *Desmostachya-Nerium oleander*-Gesellschaft abgelöst.

- Über eine in den Lebensformen vergleichbare ökologische Reihe berichten Braun-Blanquet/De Bolos (1957) aus dem Ebro-Becken. Dort steht ein Tamaricetum gallicae (= *T. canariensis*) in Kontakt zum Equiseto-Erianthetum (Imperato-Erianthion). In höheren Lagen werden die *Tamarix*-Gesellschaften dort vom Rubo-Nerietum oleandri abgelöst (De Bolos 1955).

Ein eingehender Vergleich der Standortverhältnisse, der Lebensformen und der Struktur dieser Gesellschaften erscheint lohnend.

5. Summary

Wadi-margin plant communities of the Y.A.R. are studied in the Wādī Maur region (Tihāma) and in the adjacent basin of aṭ-Ṭūr (Mountainous Tihāma). Sectional drawings (Fig. 3–9) across the wadis illustrate life-forms, stratification and zonation of the plant communities. The vegetation transects concentrate on dominant and characteristic plant species. Since they are based on the actual vegetation pattern, they also cover land use and irrigation systems.

At autochthonous water-courses in the Mountainous Tihāma, riparian forests start with broad-leaved *Combretum molle* stands in deeply incised ravines at higher elevations. On lower slopes, they are replaced by drought-deciduous *Acacia abyssinica* woodlands. In the basin itself the wadi bottom is filled with sediments. There, cobble and sandy islands are colonized by an *Acacia ehrenbergiana-Cassia senna* woodland. In the Tihāma we find only shallow hollows with episodic waterflow. A *Cadaba-Schweinfurthia pterosperma*-community is growing there.

Vegetation types on allochthonous large wadis are summarized in Fig. 10. Within the Mountainous Tihāma, mesophytic riverine forests (*Breonadia salicina*- and *Berchemia discolor*-typus) are replaced by the *Tamarix-Desmostachya bipinnata*-typus. The latter is substituted in the central part of the Tihāma by a *Tamarix-Saccharum spontaneum*-type. In the coastal part, a

Tamarix-Zygophyllum simplex-community covers saline clay in the wadi bottom.

The vegetation pattern along wadis is above all determined by drainage system (riverbed gradient, stream velocity) and grain size of the sediment. Less important are groundwater level, overflow frequency and climatic conditions.

Finally chorological phenomena are pointed out at *Tamarix*-communities of the Arabian Peninsula, North and East Africa. In tropical climate (maxima of precipitation in summer, never frost) *Tamarix*-stands are accompanied by Eritreo-Arabian *Jatropha* species, in arid to semiarid extratropical climate (maxima of precipitation in winter, frost) by Arabo-Sindian *Rhazya* species, in less arid conditions by Mediterranean-Irano-Turanian *Nerium* species.

6. Literatur

Baierle, H.-U. / El-Sheikh, A. M. / Frey, W., Vegetation und Flora im mittleren Saudi Arabien (aṭ Ṭā'if – ar-Riyāḍ). Beih. Tübinger Atlas Vorderer Orient, Reihe A (Naturwiss.) Nr. *20*, 1985, Wiesbaden.

Baum, B., The genus Tamarix, 1978, Jerusalem.

Bolos, O. de, De vegetatione notulae II, Collect. Bot. *5*, 1956, 195–268.

Bolos, O. de, Tabula vegetationis Europae occidentalis, Acta Geobot. Barcin. *3*, 1968, 5–8.

Boudouresque, E. / Schulz, E., The flora and vegetation of NE Niger (Djado, Kaouar and Ténéré), Willdenowia *11*, 1981, 363–394.

Braun-Blanquet, J. / Bolos, O. de, Les groupements végétaux du bassin moyen de l'Ebre et leurs dynamismes, An. Estac. Exper. Aula Dei *5*, 1957, 1–266.

Deil, U. / Müller-Hohenstein, K., Beiträge zur Vegetation des Jemen I – Pflanzengesellschaften und Ökotopgefüge der Gebirgstihama am Beispiel des Beckens von At Tur (J.A.R.), Phytocoenologia *13*, 1985 a, 1–102.

Deil, U. / Müller-Hohenstein, K., Die Gefäßpflanzen des Beckens von At Tur (Gebirgstihamah, Nordjemen): Häufigkeit, Standortansprüche und Vergesellschaftung, Bot. Jahrb. *106*, 1985 b, 23–52.

Escher, H., Wirtschafts- und sozialgeographische Untersuchungen in der Wadi Mawr Region (Arabische Republik Jemen), Beih. Tübinger Atlas Vorderer Orient, Reihe B (Geisteswiss.), Nr. *23*, 1976, Wiesbaden.

Flohn, H., Über die Ursachen der Aridität Nordost-Afrikas, Würzb. Geogr. Abh. *12*, 1964, 25–41.

Flohn, H., Klimaprobleme am Roten Meer, Erdkunde *19*, 1965, 179–191.

Kopp, H., Agrargeographie der arabischen Republik Jemen, Erlanger Geogr. Abh. Sonderb. *11*, 1981, Erlangen.

Kopp, H., Die Entwicklung von Viehhaltung und Holznutzung im Rahmen der Agrarstruktur des Beckens von At Tur, in: Kopp, H. (Ed.), Jemen Studien Band *1*, 1984, Wiesbaden.

Kraft, W. / Garling, F. / Matecki, J., Die hydrogeologischen Verhältnisse in der mittleren Tihama, Zeitschr. Angew. Geol. *17*, 1971, 239–249.

Meusel, H. / Jäger, E. / Weinert, E., Vergleichende Chorologie der zentraleuropäischen Flora, Band *2*, 1978, Jena.

Müller, M., Selected climatic data for a global set of standard stations for vegetation science, 1982, Den Haag.

Quèzel, P., La végétation du Sahara, 1965, Stuttgart.

Rathjens, C. / Kerner, G., Das Klima von Jemen, Deutscher Wetterdienst, Seewetteramt, Einzelveröff. *11*, 1956, 13–37, Hamburg.

Rudich, D. / Danin, A., The vegetation of the Hazeva area, Israel, Israel J. Bot. *27*, 1978, 160–176.

Täckholm, V., Student's Flora of Egypt, 1974, Beirut.

Tutin, T. et al. (Ed.), Flora Europaea, Vol. *3*, 1972, Cambridge.

Vesey-Fitzgerald, D., Vegetation of the Red Sea coast south of Jedda, Saudi Arabia, J. Ecol. *43*, 1955, 477–489.

Vesey-Fitzgerald, D., The vegetation of the Red Sea coast north of Jedda, Saudi Arabia, J. Ecol. *45*, 1957 a, 547–562.

Vesey-Fitzgerald, D., The vegetation of central and eastern Saudi Arabia, J. Ecol. *45*, 1957 b, 779–798.

Wendelberger, G., Überschwemmte Hartholzauen?, Vegetatio *28*, 1973, 253–281.

Wendelberger, G., Die Auenwälder der Donau in Österreich, Coll. Phytosoc. *9*, 1984, 19–54.

Westinga, E. / Thalen, D., A survey and problem analysis of the rangeland in the Rada district, Rada' Integrated Rural Development Project, Techn. Note *5*, 1980, Arnhem.

Zohary, M. (Ed.), Flora Palaestina, Vol. *1*, 1966, Jerusalem.

Zohary, M. / Heyn, C. C. / Heller, D., Conspectus florae orientalis, Fasc. *2*, 1983, Jerusalem.

Dank

Die Geländearbeiten und die Auswertung des Herbarmaterials fanden in Zusammenarbeit mit Herrn K. Müller-Hohenstein statt. Beim Bestimmen der Pflanzen waren die Kollegen am Royal Botanic Garden Edinburgh sehr behilflich. Die Reisen wurden finanziell von der Stiftung Volkswagenwerk, organisatorisch von den Universitäten Tübingen und Sana'a unterstützt. Frau M. Belz besorgte die Reinschrift des Manuskriptes, Frau E. Schill die Reinzeichnung der Abbildungen, Herr K. Keil führte die Reproarbeiten aus. Ihnen allen gilt mein herzlicher Dank.

Dr. Ulrich Deil, Institut für Geowissenschaften der Universität Bayreuth, Lehrstuhl für Biogeographie, Universitätsstr. 30, D-8580 Bayreuth.

Photo 1 Wādī Lā'a vor Eintritt in das Becken von aṭ-Ṭūr (mesophytischer Galeriewald am Ufer; ackerbauliche Nutzung der Hochterrassen und Unterhänge; degradierte *Acacia asak*-Trockengehölze an den Hängen).

Photo 2 Zusammenfluß von Wādī Lā'a und Wādī Maur vor Austritt aus dem Gebirge (Tamarisken-Initialen im Wadibett; *Desmostachya*-Saum und *Acacia ehrenbergiana-Cassia senna*-Trockengehölz auf sandigen Niederterrassen am Gleithang; aufgelichtetes *Acacia ehrenbergiana-Salvadora persica*-Gehölz oder ackerbauliche Nutzung auf den Ebenen; *Acacia mellifera*-Trockengehölze an den Hängen).

Photo 3 Wādī Maur bei az-Zuhra (Tamariskenbüsche und *Saccharum spontaneum*-Horste im Wadibett).

Photo 4 Nordarm des Wādī Maur (Wādī 'Abbās) nordwestlich az-Zuhra (Blick von einem sandigen Dünenzug mit *Acacia ehrenbergiana* über das Tamariskengebüsch im Wadibett; im Hintergrund saylbewässertes Land mit vereinzelten Tamarisken- und *Capparis decidua*-Büschen).

Photo 5 Blick von der südlichen Randstufe in das Becken von aṭ-Ṭūr (Mittel- und Unterlauf des Wādī Labṭah mit *Acacia ehrenbergiana-Cassia senna*-Trockengehölzen auf Schotterinseln; degradierte *Acacia mellifera*-Trokkengehölze an den Unterhängen und auf den Fußflächen im Beckeninneren).

Photo 6 Dünenfeld nördlich az-Zuhra (Blick vom Dünenkamm mit *Panicum turgidum*-Horsten und *Blyttia spiralis/Salsola spinescens*-Büschen in ein *Cadaba-Schweinfurthia*-Dünental).

Wolfgang Frey and Harald Kürschner

Masqaṭ area (Oman). Remnants of vegetation in an urban habitat

Contents

1. Introduction

Little is known about the vegetation of the Sultanate of Oman, a country which by virtue of its location on the Tropic of Cancer and its topography is one of the most varied regions of the Arabian Peninsula. The mountain system in the north, the tropically influenced mountain spurs in the south and the intervening sand areas make this transition region between the Palaeotropic and Holarctic Kingdoms particularly interesting from the viewpoint of plant geography and vegetation ecology.

In the last ten years the authorities in Oman have fortunately made an increased effort to promote scientific studies; projects such as "The Oman Flora and Fauna Surveys" in 1975 and 1977 are an example. Besides yielding a well-grounded general description of the flora, these surveys revealed a number of new taxa (Davies 1979, Forman 1980, Gamal-Eldin 1984, Miller 1984, Radcliffe-Smith 1980b, Scholz 1983, Scott 1981) and led to the publication of studies in vegetation ecology that laid the foundation for knowledge about the vegetation in this country (Mandaville 1977, 1980, Radcliffe-Smith 1980a, Sankary 1980, Sale 1980).

Detailed geobotanic analysis requiring the use of elaborate techniques (in phytosociology, chorology, ecology and cartography) constitute the basis for administrative decisions on protection, conservation, recultivation and regional landscape development. They are still an exception, however (KÜRSCHNER 1986 and the unpublished reports by LAWTON 1978 a, b, 1980). With the aid of a vegetation map the present study describes the remaining fragments of vegetation, now jeopardized by urban expansion, in the town of Masqaṭ and its surroundings. A documentation is given of the natural vegetation potential, thus facilitating decisions about future landscape development in the area.

2. Study area

The mapped area occupies the coastal region (al-Batinah) and the foothills (foothills of the central Omanian mountain region) in the eastern part of the Arabian Peninsula (58° 30′ E, 23° 35′ N, Map 1, Photo 1). It consists

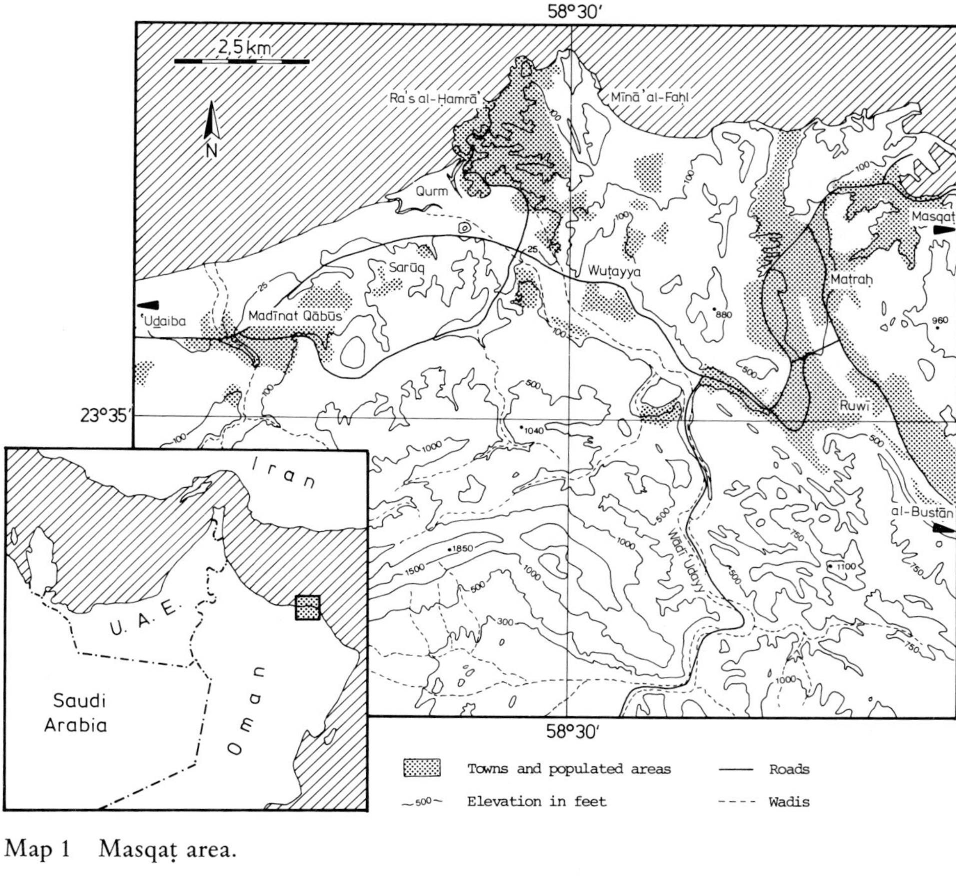

Map 1 Masqaṭ area.

of Mesozoic sediments, in part metamorphosed, uplifted and folded in tertiary times (oligocene, miocene). The presence of superposed masses of igneous rocks and cretaceous ophiolites complicate the geological situation, but in structural and petrological features this area is closely related to the southwestern Iranian mountains.

The landscape is characterized by four main physiographic features (Map 2) which are expressed also in the vegetation units.

Climatological the area belongs to the subtropical dry zone (subtropical, aride zonobiome) with winter rains, characterized by an arid desert climate with hot summers and mild winters. The mean average temperature is 27.5° C, the annual rainfall is about 106 mm (Masqaṭ, Map 2).

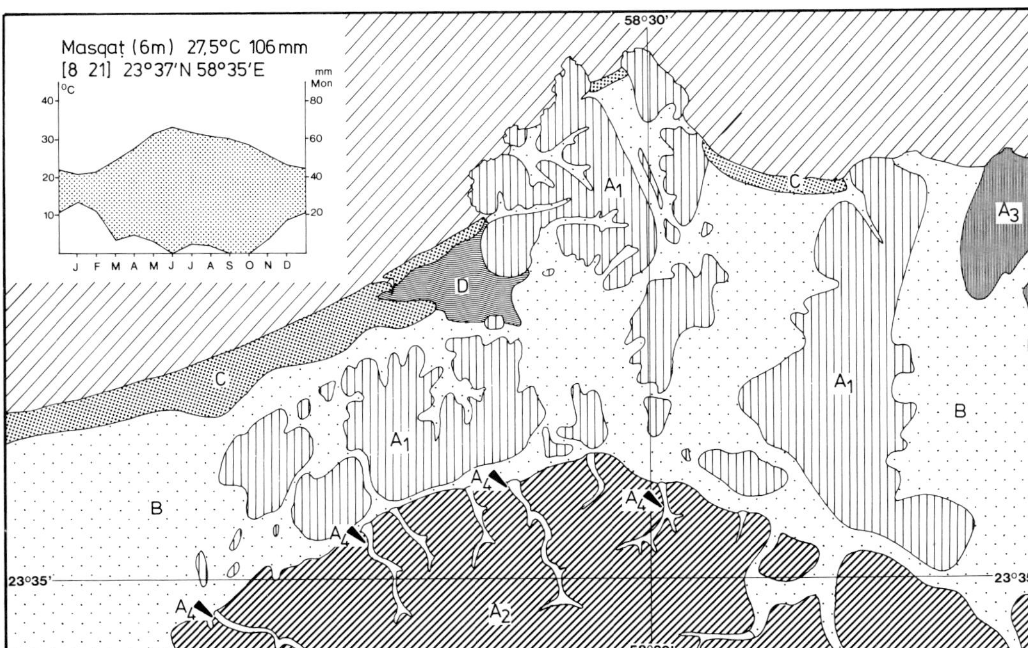

Map 2 Physiographic units and climatic conditions in the study area (explanations in the text).

A Foothills and mountain ranges (foothills of the central Omanian mountain region) characterized by scrub and dwarf-scrub.

A_1 Karstic limestones and plateaus up to 50 m (pleistocene coastal terraces).

A_2 Limestones and dolomites up to 610 m.

A_3 Rock desert (Ophiolites, Harzburgites and Peridotites).

A_4 Gorges and deep wadi cuts with wadi bottom vegetation.

B Coastal plain (al-Batinah) and wadis characterized by woodlands.

C Coastal- and eolian sand dunes. Area of psammophytic vegetation.

D Depressions with salt marshes (halophytic vegetation) and mangroves.

3. Vegetation of the Masqaṭ area (Map 3)

The vegetation map is based on the vegetation survey in February 1983 and the evaluation of aerial photographs in the scale 1 : 10 000 (aerial photography by B. K. S. Surveys Ltd., Coleraina, Co. Londonderry, N. Ireland, Sultanate of Oman, Capital area, sheet no. 6, 7, date of photography: June 1977). The classification follows the physiognomic-ecological proposal for Southwest Asia by FREY/PROBST (1977).

3.1 Woodlands

Mangrove forests **(1.1,** *Avicennia marina* community type ZAHRAN 1982, *A. marina* community KASSAS/ZAHRAN 1967, BATANOUNY 1981, *A. marina* association FREY et al. 1984)

The mangrove forests occur in the former delta of Wādī ʿUdayy (Qurm, Map 1). The estuary here was created by sandbanks and is connected to the sea by two tidal channels (outflow of the Wādī ʿUdayy). Through subterranean inflow of ground water, fresh water is mixed with sea water (mixohaline mangrove, Cl^- 2.0–2.3 %, KÜRSCHNER 1986) in the rhythm of the tides which allows the occurrence of mangrove forests (pure stands of *Avicennia marina*). They are characteristic of the tidal channels, runnels and gullies permanent flooded or overflooded at high tide (on lagoonal silts and mudbanks; muddy, loamy-sandy, wet and moderately saline soils which are poor in oxygen and rich in organic substances; compare KÜRSCHNER 1986). These mangroves are important productive biotopes in the Masqaṭ area as first link in the detritus food chain of marine organisms which give nourishment to numerous inhabitants of the estuary. Among other factors this "food trap" is responsible for the wealth of fish near Ra's al-Ḥamrā' and has beside his ecologic also an economic value.

Mixed formation of *Mangrove forests* (1.1) and *Halophytic semi-woody shrub communities* (14.1)

Mangrove forests, densely interspersed by halophytes are characteristic of the sandy central part of the former delta of Wādī ʿUdayy (Qurm, Map 1). From the tidal channels to the sand dunes this mixed formation shows a clear zonation attributed to edaphical parameters (Tab. 1). Increasing accumulation of sand in the central part leads to dessication and sanding up of the mangroves.

Sample record: Qurm Nature Reserve, eolian sands and sabkhat deposits; alt.: 5–10 m; coverage: up to 40 %; date: 17 Febr. 1983.

Trees and shrubs: *Avicennia marina, Indigofera oblongifolia, Suaeda aegyptiaca.*

	Avicennia marina zone		*Halopeplis perfoliata-Aeluropus lagopoides* zone		*Indigofera oblongi-folia-Suaeda aegyp-tiaca* zone
Moisture (%):	16.0–22.5	►	12.8–16.0	►	0.9–4.9
Cl^- mg/100 g*:	716 (430)	◄	1480 (640)	►	7–50 (40–315)
$CaCO_3$ (%):	23 (23)	►	12 (14)	◄	20–28 (23–25)
Relief:	Runnels & gullies		Border of runnels		Sand dunes & small elevations

◄ increase *surface values, values at 50 cm depth in brackets

Tab. 1 Zonation of vegetation dependent on edaphical parameters at the central part of Qurm (after Kürschner 1986).

Chamaephytes: *Halopeplis perfoliata, Suaeda fruticosa, S. vermiculata.*
Hemicryptophytes: *Aeluropus lagopoides, Cressa cretica, Sporobolus arabicus, S. spicatus.*

Thorn woodlands **(5.2,** *Acacia tortilis-Prosopis cineraria* community type, Photos 2, 3)

Formerly large areas of the coastal plain (al-Batinah) and the alluvial lowland of Oman were covered by xeromorphic woodlands (pseudosavannas, *Acacia* desert parkland Mandaville 1977), highly influenced and dependent on groundwater. These *Thorn woodlands*, originally consisting of *Acacia ehrenbergiana, A. tortilis, Prosopis cineraria* and *Ziziphus spina-christi* [Acacietea flavae-iranica Zohary 1973, *A. flava-Prosopis spicigera* (Syn. *P. cineraria*) association Zohary 1963], are found in remnants throughout the study area. According to the underground and the anthropozoogenetic influence they appear in different degradation stages, ranging from well-developed woodland with rich shrub and herb layers (today very rare) to completely destroyed woodland consisting of single trees only (mainly in the Wādī ʿUdayy, Wādī Ruwī and Wādī Maṭraḥ).
In the Masqaṭ area, five varieties of the *Thorn woodland* are distinguished (5.2 a–5.2 e, Map 3).

Variety 1 (5.2 a)

This variety occurs on sand-covered alluvium and fluviatile gravels in small wadis which have been subjected to little human influence. In these fragments of xeromorphic woodlands all the characteristic species are still present: *Acacia tortilis, Maerua crassifolia, Prosopis cineraria* and *Ziziphus spina-christi.* They colonize the small wadi cuts between Ra's al-Ḥamrā' and Mīnā' al-Faḥl and the wadis west of Madīnat Qābūs.
Sample record: Wadi southwest of Madīnat Qābūs, on fluviatile gravel; alt.: 30 m; coverage: 20 %; date: 20 Febr. 1983.

Trees: *Acacia tortilis, Prosopis cineraria, Ziziphus spina-christi.*
Chamaephytes: *Euphorbia larica, Jaubertia aucheri, Leucas inflata.*
Hemicryptophytes: *Aerva javanica, Boerhavia elegans, Cenchrus pennisetiformis, Fagonia bruguieri, F. indica, Morettia parviflora, M. canescens, Physorrhynchus chamaerapistrum, Rhazya stricta, Tephrosia apollinea.*
Therophytes: *Aristida mutabilis, Arnebia hispidissima, Enneapogon desvauxii, Euphorbia granulata, Launaea intybacea, Oligomeris linifolia, Plantago ovata, Rumex vesicarius, Schismus barbatus, Seetzenia lanata, Senecio flavus.*
Geophytes: *Asphodelus fistulosus* var. *tenuifolius.*

Variety 2 (5.2b)

This variety represents the annual-rich formations of the *Thorn-woodlands.* Growing on sand-covered alluvium, it is found only at occasional sites such as Wādī Sarūq. After spring rains there is an abundance of annuals in the undergrowth.
Sample record: Wādī Sarūq, alluvial gravel and fluviatile sands; alt.: 15 m; coverage: tree and shrub layer 30–40%, herb layer 30%; date: 15 Febr. 1983.
Trees and shrubs: *Acacia tortilis, Lycium shawii, Ochradenus arabicus.*
Chamaephytes: *Cornulaca monacantha, Indigofera arabica, I. intricata.*
Hemicryptophytes: *Argyrolobium roseum, Cassia italica, Convolvulus virgatus, Cyperus conglomeratus, Fagonia indica, Panicum turgidum, Pergularia tomentosa, Tephrosia apollinea, T. nubica, T. purpurea.*
Therophytes: *Amaranthus graecizans, Aristida mutabilis, Arnebia hispidissima, Astragalus eremophilus, A. vogelii* ssp. *fatimensis, Centaurea pseudosinaica, Chrozophora scabulosa, Dactyloctenium aristatum, Eragrostis cilianensis, Erodium laciniatum* var. *pulverulentum, Euphorbia granulata, Launaea capitata, Lotus halophilus, Malva parviflora, Neurada procumbens, Paronychia arabica, Plantago ovata, Schismus barbatus, Tribulus parvispinos.*

Variety 3 (5.2c, Photo 2)

Variety 3 comprises the best-preserved and densest pseudosavannas. In the area of the town these forest remnants of the Batinah are usually found only in fragments along wadi banks (e.g. Wādī Sarūq, Wādī Wuṭayya), where transported fine material has been deposited. Relatively large, dense and undisturbed stands are found in the direction of 'Uḏaiba. On species composition compare variety 2 (5.2b).

Variety 4 (5.2d, Photo 3)

As in variety 3, but more open and poorer in species as a result of more severe anthropozoogenic disturbance. Today this variety is most common in Masqaṭ. The acacias appear to have a random distribution but actually tend to grow along the runoff channels and depressions (distribution is dependent on groundwater).

Variety 5 (5.2 e)

This variety consists of *Acacia tortilis* stands on talus, foothill pediments and other moist habitats near groundwater along the eastern hill ranges. The acacias are usually free-standing. In the undergrowth there are also taxa of the scrub formations (8.2).
On species composition compare variety 2 (5.2 b) and 8.2 a, c.

3.2 Scrub

Xeromorphic, very open scrub (semidesert shrublands) (8.2, *Commiphora myrrha-Euphorbia larica* community type)

The lower hills and the slopes of the Pleistocene coastal terraces of the Batinah are scattered with remnants of a scrub formation (dry deciduous, microphyllous woodlands, *Euphorbia larica* shrub slopes MANDAVILLE 1977) which is typical of the lower hill zone of Oman and comparable with those of the southern Iranian coastal region (Larestan, Makran; Euphorbietea laricae ZOHARY 1973, Acacieto-Euphorbietum SANKARY 1980). Characteristic taxa are *Acacia etbaica*, *Commiphora myrrha*, *Grewia tenax* var. *erythraea*, *Euphorbia larica* and *Ochradenus arabicus* (*Commiphora myrrha-Euphorbia larica* community type of the Euphorbietea laricae). These species settle on a very rugged relief with rapid water run-off and pronounced erosion (karstic rocks). At present these shrublands are frequently devastated and they are in danger of disappearing as a result of increasing building activity. Depending on location, on edaphic and microclimatic conditions, and on degree of destruction, four states can be distinguished.

Variety 1 (8.2 a)

This variety occurs on the karstic slopes of the plateaulike former coastal terrace, which rises to a height of up to 50 m. It is particularly endangered by the expansion of the urban area. The typical plants of the dry deciduous microphyllous woodlands are completely devastated and seldom reach a height of more than 1 m.
Sample record: Wādī Sarūq, on karstic rocks (limestones); alt.: 15–50 m; coverage: 20 %; date: 15 Febr. 1983.
Shrubs: *Commiphora myrrha, Euphorbia larica, Grewia tenax* var. *erythraea, Lycium shawii, Maerua crassifolia, Ochradenus arabicus.*
Chamaephytes: *Andrachne aspera, Halothamnus bottae, Iphiona scabra, Leucas inflata, Seddera latifolia, Vernonia arabica.*

Hemicryptophytes: *Aerva javanica, Argyrolobium roseum, Blepharis ciliaris, Cenchrus pennisetiformis, Chrysopogon plumosus, Cleome brachycarpa, Crotalaria aegyptiaca, Cymbopogon commutatus, Dichanthium foveolatum, Fagonia bruguieri, Kohautia retrorsa, Morettia parviflora, Taverniera glabra, Tetrapogon villosus.*

Therophytes: *Cometes surattensis, Launaea intybacea, L. mucronata, Plantago ovata.*

Variety 2 (8.2b)

This formation is confined to the narrow and in some cases deeply incised wadi outlets and gullies of the foothills and mountain ranges (mountain wadi associations, MANDAVILLE 1977). It differs from the preceeding variety (8.2a) in that *Dyerophytum indicum* is co-dominant. This species does not occur in the other shrubland formations. Owing to better ecological conditions (more moisture, higher humidity, shade which reduces the evaporation rate) this wadi bottom vegetation is much denser and more vigorous (*Euphorbia larica* attains heights of up to 2 m here). It also has a higher coverage and possesses a well-developed herb layer dominated by tussock grasses. Above the high-water line single large individuals of *Ficus salicifolia* and *Ziziphus spina-christi* occur.

Sample record: End of Wādī Sarūq, deeply incised wadi outlets; alt.: 100 m; coverage: up to 40%; date: 15 Febr. 1983.

Trees and shrubs: *Acacia etbaica, Commiphora myrrha, Dyerophytum indicum, Ephedra ciliata, Euphorbia larica, Ficus salicifolia, Lycium shawii, Ochradenus arabicus, O. aucheri, Periploca aphylla, Ziziphus spina-christi.*

Chamaephytes: *Andrachne aspera, Hibiscus micranthus, Iphiona scabra, Jaubertia aucheri, Lavandula subnuda, Leucas inflata, Pseudogaillonia hymenostephana, Pulicaria glutinosa.*

Hemicryptophytes: *Cleome rupicola, Commicarpus boissieri, Convolvulus virgatus, Enneapogon schimperanus, Farsetia stylosa, Forsskaolea tenacissima, Haplophyllum tuberculatum, Heliotropium calcareum, Hyparrhenia hirta, Kohautia caespitosa, Launaea spinosa, Morettia parviflora, Polygala erioptera, Salvia aegyptiaca, Schweinfurthia papilionacea, Taverniera glabra, Tricholaena leucantha.*

Therophytes: *Cometes surattensis, Galium setaceum* ssp. *decaisnei, Seetzenia lanata, Senecio flavus.*

Variety 3 (8.2c)

Variety 3 characterizes the spur slopes between 50 and 300 m in the central mountain region of Oman. Between Ruwī (al-Bustān) and Madīnat Qābūs it is of almost uniform composition. Here a number of taxa are dominant that also occur in the mountains of southern Iran. They are evidence of the close relationship between these areas in floristic history. Examples are *Euphorbia larica, Helichrysum makranicum, Jaubertia aucheri,*

Pseudogaillonia hymenostephana, Pseudolotus makranicus, Teucrium stocksianum and *Viola cinerea*. Together with *Dyerophytum indicum, Acridocarpus orientalis* and *Reptonia mascatensis*, the latter two of which are restricted to the upper parts of Ǧabal al-Aḫḍar, these species confirm ZOHARY's (1963) view that this is a major relict area of the mesic Afro-Malesian flora that was later replaced by a Tertiary xero-tropical flora and gave rise to numerous endemics.

Sample record: Mountain range south of Madīnat Qābūs, mainly karstic limestones and dolomites; alt.: 50–300 m; coverage: 20 %; date: 20 Febr. 1983.

Shrubs: *Acacia tortilis, Capparis cartilaginea, C. spinosa* var. *mucronifolia, Ephedra ciliata, Euphorbia larica, Grewia tenax* var. *erythraea, Lycium shawii, Maerua crassifolia, Ochradenus arabicus, O. aucheri, Periploca aphylla, Rhus aucheri, Salvadora persica.*

Chamaephytes: *Andrachne aspera, Hibiscus micranthus, Iphiona scabra, Jaubertia aucheri, Lavandula subnuda, Leucas inflata, Pseudogaillonia hymenostephana, Pulicaria glutinosa, Teucrium stocksianum, Vernonia arabica.*

Hemicryptophytes: *Adenosciadium arabicum, Aerva javanica, Boerhavia diffusa, Caralluma aucheriana, Cenchrus pennisetiformis, Cheilanthes catanensis, Cleome brachycarpa, C. droserifolia, C. rupicola, Commicarpus boissieri, Convolvulus virgatus, Cymbopogon commutatus, Cyperus glaber, Diplotaxis harra, Fagonia bruguieri, F. indica, Farsetia stylosa, Forsskaolea tenacissima, Glossonema varians, Haplophyllum tuberculatum, Helichrysum makranicum, Heliotropium calcareum, Hochstetteria schimperi, Morettia canescens, Phagnalon viridifolium, Polygala erioptera, Pseudolotus makranicus, Pycnocycla aucherana, Reseda aucheri, Viola cinerea.*

Therophytes: *Aristida adscensionis, Cometes surattensis, Galium setaceum* ssp. *decaisnei, Herniaria mascatensis, Senecio flavus.*

Variety 4 (8.2 d)

This variety, which has chenopods in the dwarf-shrub layer, occurs near the coast between Ra's al-Ḥamrā' and Mīnā' al-Faḥl. As a result of eolian transport and deposition of salts there is slight salinity, which favours the occurrence of halophilous and salt-tolerant species (spray and dust halophytes in the wave and reef zone).

The species composition is as in variety 1 (8.2 a), but *Anabasis setifera, Halothamnus bottae, Hammada salicornica* and *Salsola rubescens* are present.

3.3 Dwarf-scrub and related communities

Xeromorphic dwarf-shrubland with trees and shrubs **(9.41,** *Moringa peregrina* community type, Photo 4)

At an altitude of 300 m the *Xeromorphic, very open scrub* (8.2 c) of the lower hills merges into a formation characterized by *Moringa peregrina*

(*Moringa peregrina* community type). There are large stretches where this is the only arboreal species. These trees, up to 3 m high, characterize the area of Ǧabal Muraywaḥ, situated south-east of Madīnat Qābūs. It is not clear whether the lower limit of distribution is caused by dryness or whether there are anthropozoogenic causes.

Sample record: Ǧabal Muraywaḥ, severely weathered dolomites; alt.: 300–360 m; coverage: tree layer less than 5%, herb layer 30%; date: 21 Febr. 1983.

Trees: *Moringa peregrina.*

Shrubs: *Euphorbia larica, Grewia tenax* var. *erythraea, Maerua crassifolia, Ochradenus arabicus.*

Chamaephytes: compare 8.2c, additionals: *Helianthemum lippii, Indigofera arabica.*

Hemicryptophytes & Therophytes: compare 8.2c.

Halophytic semi-woody shrub communities (14.1)

Adjacent to the mangroves in the outflow of Wādī ʿUdayy at Qurm there are Nubo-Sindian influenced halophyte communities characterizing the sabkha deposits. Following a classification on the basis of the dominance principle (compare KASSAS/ZAHRAN 1967, BATANOUNY 1981), this formation comprises seven community types (see KÜRSCHNER 1986, FREY/KÜRSCHNER 1986). The most characteristic and widely distributed ones are the pure stands of *Suaeda fruticosa* (*S. fruticosa* association KÜRSCHNER 1986) on elevated sandy areas and the *Halopeplis perfoliata* association (*H. perfoliata* community type ZAHRAN 1982, *H. perfoliata* community KASSAS/ZAHRAN 1967, BATANOUNY 1981), which is typically found along the wet and strongly saline border of the tidal channels and runnels.

Sample record: Qurm Nature Reserve, eolian sands and sabkha deposits; alt.: 0–5 m; coverage: up to 90%; date: 17 Febr. 1983.

Shrubs: *Arthrocnemum macrostachyum, Atriplex leucoclada, Indigofera oblongifolia, Tamarix ramosissima.*

Chamaephytes: *Halopeplis perfoliata, Heliotropium kotschyi, Limonium stocksii, Suaeda aegyptiaca, S. fruticosa, Zygophyllum qatarense.*

Hemicryptophytes: *Aeluropus lagopoides, Anabasis setifera, Cressa cretica, Juncus rigidus, Sporobolus arabicus, S. spicatus.*

3.4 Deserts and other scarcely vegetated areas

Scarcely vegetated sand-dunes (17.1, Photo 5)

South-west of Madīnat Qābūs the relief is formed by up to 40 m-high coastal dunes and former inland dunes. These old dunes are well stabilized and are occupied by occasional shrubs and numerous psammophytes.

The characteristic species here is *Cyperus conglomeratus* (formation 17.1 a). Another smaller area of these old dunes is located at the end of Wādī Sarūq (south-east of Madīnat Qābūs, left side of the valley). In this landscape, where some of the dunes are very well stabilized, *Hyparrhenia hirta* is dominant (formation 17.1 b).

Sample record: Dunes south-west of Madīnat Qābūs, Tertiary coastal dunes (17.1 a); alt.: 40 m; coverage: less than 1 %; date: 20 Febr. 1983.

Shrubs: *Calligonum comosum, Leptadenia pyrotechnica.*

Chamaephytes: *Crotalaria aegyptiaca, Cyperus conglomeratus, Dipterygium glaucum, Heliotropium kotschyi, Panicum turgidum.*

Therophytes: *Aristida adscensionis, Monsonia heliotropioides, Tragus berteronianus.*

Sample record: Dunes south-east of Madīnat Qābūs, Tertiary dunes (17.1 b); alt.: 15–40 m; coverage: up to 10 %; date: 15 Febr. 1983.

Chamaephytes: *Cornulaca monacantha.*

Hemicryptophytes: *Aerva javanica, Crotalaria aegyptiaca, Cyperus conglomeratus* var. *aucheri, Dichanthium foveolatum, Dipterygium glaucum, Hyparrhenia hirta, Panicum turgidum, Pennisetum divisum, Tephrosia quartiniana.*

Therophytes: *Arnebia hispidissima, Rumex vesicarius.*

Coastal dunes (17.12, *Halopyrum mucronatum* community type, *Suaeda fruticosa-S. vermiculata* community type Zahran 1982)

Coastal dunes consisting of loose or slightly fixed white sands are found along the whole marine border of the study area (compare also 17.12, 14.). As they have been closely examined only in the Qurm Nature Reserve (Kürschner 1986), they were mapped as a separate formation only in this area.

Depending on the mobility, texture and exposure of the marine and eolian deposits one can distinguish several different community types of different distribution and importance. The factors responsible for their patterns are complex and not yet well understood. The most important types are: the *Halopyrum mucronatum* community type Zahran 1982 (*H. mucronatum* community Kassas/Zahran 1967, Batanouny 1981), a sand-stabilizing community on deep and loose sands that forms the first vegetation belt following the beach zone. It merges abruptly into the *Suaeda fruticosa-S. vermiculata* community type Zahran 1982 (*S. vermiculata* community Batanouny 1981) which is typical of poorly to moderately fixed higher dunes toward the land. Small depressions in this belt are dominated by *Lotus garcinii* (see Kürschner 1986).

For species composition compare 17.12, 14.

Mixed formation of *Coastal dunes* (17.12) and *Herbaceous and semi-woody salt swamps* (14.)

Partly stabilized large dunes intermixed with deep, loose, eolian sands strongly affected by wind erosion are typical of the coastal region west of the Qurm Nature Reserve. In the interior this dune landscape is characterized by *Calligonum comosum* and *Cornulaca monacantha*, while near the coast the *Suaeda fruticosa-S. vermiculata* community type is very common (compare 17.12). In depressions, hollows and channels halophytes (especially *Aeluropus lagopoides*) make up the typical vegetation.
Sample record: Dunes west of the Qurm Nature Reserve, eolian sands and coastal dunes; alt.: 3–10 m; coverage: less than 10%; date: 16 Febr. 1983.
Shrubs: *Atriplex leucoclada, Tamarix ramosissima.*
Chamaephytes: *Cornulaca monacantha, Limonium stocksii, Sphaerocoma aucheri, Suaeda fruticosa, S. vermiculata.*
Hemicryptophytes: *Aeluropus lagopoides, Cistanche phelypaea, Crotalaria aegyptiaca, Cyperus conglomeratus, C. kali, Dipterygium glaucum, Halopyrum mucronatum, Lotus garcinii, Sporobolus spicatus.*
Therophytes: *Cornulaca aucheri.*

Rock desert **(18.31,** thinly scattered with dwarf-shrubs)

The plateaulike erosion surface of the up to 50 m-high Pleistocene coastal terrace has only a sparse cover of vegetation. Owing to strong wind erosion and dessication, the scrub here is unable to thrive; only a few dwarf-shrubs and hemicryptophytes remain.
Sample record: South of Sarūq, karstic limestone plateau; alt.: 50 m; coverage: less than 5%; date: 15 Febr. 1983.
Chamaephytes: *Indigofera arabica, Iphiona scabra, Leucas inflata, Psdeudogaillonia hymenostephana.*
Hemicryptophytes: *Blepharis ciliaris, Boerhavia elegans, Cenchrus pennisetiformis, Chrysopogon plumosus, Dichanthium foveolatum, Enneapogon schimperanus, Fagonia bruguieri, Morettia parviflora, Launaea spinosa.*
Therophytes: *Aristida abnormis, Launaea mucronata.*

Rock desert **(18.31)**

On account of a poor water supply the extremely karstic erosion surface of the mountain ranges adjacent to the south has only a sparse cover of dwarf-shrubs at elevations between 120 and 160 m. Dense vegetation with shrubs *(Euphorbia larica)* occurs only in erosion channels. On species composition compare 8.2 c.

Debris and gravel desert **(18.32,** Photo 6)

Extensive debris and gravel deserts are found between Maṭraḥ and Masqaṭ and along Wādī Ruwī as far as al-Bustān (eastern wadi limit, 18.32 a).

Geologically they consist of Cretaceous ophiolites, harzburgites and peridotites which are subjected to strong physical weathering (thermal fracture). This loose material is almost devoid of phanerogams, and the sparse plant cover is generally confined to the lower parts of the talus fans in Wādī Ruwī (important species are *Boerhavia elegans*, *Gypsophila bellidifolia*, *Polycarpaea spicata* and *Stipagrostis hirtigluma*). The most conspicuous element of these talus fans is *Salvia macilentha* (chorotype: Nubo-Sindian, endemic with strong disjunction Oman-Iran), which is limited to these sites in the study area.

The numerous small talus fans in the area characterized by limestones and dolomites were mapped only at one sample location in the western part of the map (18.32b). Owing to mobility and continuous addition of detritus they too are almost free of vegetation.

Sample record: Wādī Kabīr, east of Ruwī, Cretaceous Ophiolites; alt.: 100–150 m; coverage: less than 1%; date: 22 Febr. 1983.

Hemicryptophytes: *Argyrolobium roseum, Boerhavia elegans, Cleome droserifolia, Dichanthium foveolatum, Gypsophila bellidifolia, Misopates orontium, Morettia parviflora, Salvia macilentha, Tephrosia apollinea, Tricholaena leucantha.*

Therophytes: *Aristida mutabilis, Euphorbia arabica, Polycarpaea spicata, Rumex vesicarius, Stipagrostis hirtigluma.*

Severely degraded areas

The area most severely affected by urban expansion are the large wadi systems of Wādī ʿUdayy, Wādī Wuṭayya (including the coastal region east of Mīnā' al-Faḥl) and Wādī Ruwī (Wādī Kabīr). Owing to their lack of relief they are heavily used for infrastructure, and since groundwater is relatively near the surface in these drainage systems there is a steady increase on construction activity including new housing developments.

The trees originally found in these wadis (*Acacia ehrenbergiana*, *A. tortilis*, *Prosopis cineraria*, *Ziziphus spina-christi*; *Acacia tortilis-Prosopis cineraria* community type) have disappeared except for a few large individual specimens, and ruderal communities have replaced them. Large areas are nearly plantless; they are highly contaminated (refuse and rubble heaps, storage areas) and irreversibely destroyed.

Sample record: Wādī Wuṭayya, alluvial gravel; alt.: 15 m; coverage: 10–20%; date: 14 Febr. 1983, fragments of the natural wadi vegetation, strong ruderal influence.

Trees: *Acacia tortilis, Prosopis cineraria, Ziziphus spina-christi.*

Hemicryptophytes: *Aerva javanica, Boerhavia diffusa, Cenchrus pennisetiformis, Farsetia aegyptiaca, Morettia parviflora, Osteospermum vaillantii, Physorrhynchus chamaerapistrum, Tephrosia apollinea.*

Therophytes: *Aristida adscensionis, Chenopodium murale, Eragrostis barrelieri, E. cilianensis, E. ciliaris, Launaea mucronata, Oligomeris linifolia, Tribulus parvispinos, Zygophyllum simplex.*

Sample record: Wādī Sarūq, alluvial gravel; alt.: 15 m; coverage: up to 80%; date: 15 Febr. 1983, watering place with hydrophytic bank vegetation, strong ruderal influence (remnants of hydrophytic vegetation).

Trees: *Acacia tortilis, Prosopis cineraria, P. juliflora* (cult.), *Ziziphus spina-christi.*

Hemicryptophytes: *Chloris barbata, Lippia nodiflora, Pennisetum divisum, Tricholaena leucantha.*

3.5 Climax vegetation and substitutes

Given the formations existing in the area and the various observed stages of degradation the following anthropozoogenic regressive successions are assumed to occur:

a) Wadis of the coastal plain

Theoretical climax vegetation

Acacia ehrenbergiana-Prosopis cineraria-Ziziphus spina-christi woodlands (*Acacia flava-Prosopis spicigera* assoc. ZOHARY 1963)

▽

Remnants of the *Acacia-Prosopis* assoc., strong ruderal influence

▽

Ruderal wadi formations without trees dominant

Substitute communities

b) Coastal plain (al-Batinah)

Theoretical climax vegetation

Relatively dense *Acacia tortilis* woodlands (Acacietea tortilis subsudanica ZOHARY 1973, *Acacia tortilis* community BATANOUNY 1981)

▽

Devastated open *Acacia tortilis* woodlands

▽

Acacia tortilis shrublands and dwarf-scrub

▽

Herbaceous formations (ephemeral communities) and eroded sandy plains

Substitute communities

c) Foothills and mountain ranges

Theoretical climax vegetation ↕ Substitute communities

Commiphora myrrha-Euphorbia larica shrublands (Dry deciduous, microphyllous woodlands; Euphorbietea laricae ZOHARY 1973)

▽

Remnants of the *Commiphora-Euphorbia* shrublands

▽

Dwarf-shrub formations with *Euphorbia larica*

▽

Rock desert

It must be noted that there remain only a few places in the study area where vegetation units can be found with a natural composition. Usually the units are highly degraded in parts (2nd degree substitute communities) and have an altered species composition especially in the undergrowth. With increasing urbanization and industrial development it is to be feared that within a few years the vegetation in this area will consist only of replacement communities of the final stage.
To make possible a progressive succession and to preserve the remains of the potential natural vegetation it will be necessary to carry out a reasonable partition of the landscape that does justice to the conflicting interests of economy and ecology.

We therefore suggest:

- that several wadi systems should be placed under protection. Particularly suitable are Wādī Sarūq and its continuation into the mountains, since almost all of the above-described typical vegetation units are still present here.
- that the slopes and limestone plateaus of the Pleistocene coastal terraces with their characteristic shrubland fragments should be protected or incorporated into existing reserves (compare KÜRSCHNER 1986).
- that the coastal dunes and herbaceous and semi-woody salt swamps of the coastal region should be placed under protection in order to put a check on the increasing construction activity and prevent use of the beach zone.
- that grazing should be prevented in the sand dunes south-west of Madīnat Qābūs in order to preserve these unique vegetation units in the Masqaṭ area.

Through the establishment of the proposed reserves it would be possible to preserve the diverse vegetation of the Masqaṭ area, which is threatened with near destruction if measures are not taken to mitigate the effects of further development. It is to be hoped that the responsible authorities will take steps to maintain the diversity of vegetation in this area.

4. References

Batanouny, K. H., Ecology and flora of Qatar, Univ. of Qatar, 1981.

Davies, F. G., A new species of *Launaea* (Compositae) from Oman, Kew Bull. *33*, 1979, 294.

Forman, L. L., *Cocculus balfourii* (Menispermaceae) in Oman (Dhofar), Kew Bull. *35*, 1980, 379–381.

Frey. W. / Kürschner. H., Gulf of Aqaba, Gulf of Oman. Mangroves and coastal salines, map A VI 10.8 Tübinger Atlas des Vorderen Orients, 1986, Wiesbaden.

Frey, W. / Kürschner, H. / El-Sheikh, A. M. / Migahid, A. M., Zonation and photosynthetic pathways of halophytes along a transect from the Red Sea coast near Tawwal (Saudi Arabia), Notes R. B. G. Edinburgh *42*, 1984, 45–56.

Frey, W. / Probst, W., Classification and mapping of vegetation in the Tübinger Atlas des Vorderen Orients (TAVO) and in the supplements to the atlas, Beih. Tübinger Atlas Vorderer Orient, Reihe A (Naturwiss.) Nr. *1*, 1977, Wiesbaden.

Gamal-Eldin, E., Studies in the flora of Arabia VIII. A new *Pulicaria* from Oman, Notes R. B. G. Edinburgh *41*, 1984, 467–471.

Kassas, M. / Zahran, M. A., On the ecology of the Red Sea littoral salt marsh, Egypt. Ecol. Monogr. *37*, 1967, 297–316.

Kürschner, H., A study on the vegetation of the Qurm Nature Reserve (Muscat area, Oman), Arab Gulf J. Sci. Res. *4*, 1986 (in press).

Lawton, R. M., A reconnaissance survey of the Jebel Qara grazing land ecosystem, with particular reference to the impact of development, Min. Overseas Development 1978 a.

Lawton, R. M., Vegetation reconnaissance survey for the *Oryx* project. Report to the Sultanat of Oman and the World Wildlife Fund, Min. Overseas Development 1978 b.

Lawton, R. M., The forest potential of the Sultanate of Oman, Overseas Development Administration, Land Resources Development Centre Project Report *97* (OMAN-03-1/REP 97/80), 1980.

Mandaville, J. P., Plants. In: The Scientific Results of the Oman Flora and Fauna Survey 1975, Journ. Oman Studies Spec. Rep. *1*, 1977, 229–267.

Mandaville, J. P., *Frankincense (Boswellia)* in Dhofar. In: The Scientific Results of the Oman Flora and Fauna Survey 1977 (Dhofar), Journ. Oman Studies Spec. Rep. *2*, 1980, 87–89.

Miller, A. G., A revision of *Ochradenus,* Notes R. B. G. Edinburgh *41*, 1984, 491–504.

Radcliffe-Smith, A., The vegetation of Dhofar. In: The Scientific Results of the Oman Flora and Fauna Survey 1977 (Dhofar), Journ. Oman Studies Spec. Rep. *2*, 1980 a, 59–86.

RADCLIFFE-SMITH, A., A new species of *Jatropha* from Dhofar, Kew Bull. *35*, 1980b, 253–255.

SALE, J. B., The ecology of the mountain region of Dhofar. In: The Scientific Results of the Oman Flora and Fauna Survey 1977 (Dhofar), Journ. Oman Studies Spec. Rep. *2*, 1980, 25–54.

SANKARY, M. N., Ecology and phytogeography of Jabal al-Akdar, 20th Science Week, 1980, Damascus.

SCHOLZ, H., *Stipagrostis masirahensis* sp. nov. aus Oman (Arabien), Willdenowia *13*, 1983, 389–391.

SCOTT, A. J., A new *Suaeda* from Dhofar, Kew Bull. *36*, 1981, 558.

ZAHRAN, M. A., Ecology of the halophytic vegetation of Egypt. In: SEN, D. N. / RAJPUROHIT, K. S. (eds.), Contribution to the ecology of halophytes pp. 3–20, 1982, The Hague – Boston – London.

ZOHARY, M., On the geobotanical structure of Iran, Bull. Res. Counc. Israel, Sect. D, Botany *11*, 1963, 65–113.

ZOHARY, M., Geobotanical foundations of the Middle East, 1973, Stuttgart – Amsterdam.

Acknowledgements

The authors are grateful to His Highness Sayyid Faisal Bin Ali Al Said (Minister of National Heritage & Culture) and to Mr A. Shanfari (Director of Antiquities) for the possibility to undertake research studies in Oman. We are also grateful to Mr M. D. Gallagher (Advisor for the Oman Natural History Museum) for information and advice, to Miss R. A. King, Mr I. C. Hedge and Mr A. G. Miller (Royal Botanic Garden Edinburgh) for their help in identification of some critical specimens and to Prof Dr H. Scholz (Berlin) for identifying the grasses. Special thanks belong to Dr Dr H.-P. Uerpmann (Tübingen) for inviting us to this research journey, for help in the field work and living facilities in Muscat.
The study was supported by the Ministry of National Heritage & Culture (Oman) and the Deutsche Forschungsgemeinschaft, Sonderforschungsbereich 19 "Tübinger Atlas des Vorderen Orients".

Prof. Dr. Wolfgang Frey, Dr. Harald Kürschner, Institut für Systematische Botanik und Pflanzengeographie, Freie Universität Berlin, Altensteinstr. 6, D-1000 Berlin 33.

Photo 1 Masqaṭ area (study area), overview. On the left Madīnat Qābūs, in the middle, parts of the Pleistocenic coastal terraces, in the background the Qurm Nature Reserve and Ra's al-Ḥamrā'.

Photo 2 Remnants of well developed *Thorn woodlands* (5.2 c) on sandy alluvium near Sarūq.

Photo 3 Remnants of *Thorn woodlands* (5.2 d) south-west of Madīnat Qābūs.

Photo 4 *Xeromorphic dwarf-shrubland with trees and shrubs* (9.41, *Moringa peregrina* community type) on the mountains south-east of Madīnat Qābūs.

Photo 5 *Scarcely vegetated sand-dunes* (17.1). Stabilized old dunes south-west of Madīnat Qābūs.

Photo 6 Wādī Kabīr (Ruwī). Urban area with devastated wadi bottom vegetation. On the right *Debris and gravel desert* (18.32, Ophiolites), on the left calcareous rock with *Xeromorphic, very open scrub* (8.2 c).